How We Got From There To Here:
A Story of Real Analysis

How We Got From There To Here: A Story of Real Analysis

Robert Rogers

SUNY, Fredonia
Fredonia, New York, USA

Eugene C. Boman

Penn State, Harrisburg campus
Harrisburg, PA, USA

Edition: 2.0.2

[1] creativecommons.org/licenses/by-nc-sa/4.0/

Acknowledgements

Mactutor While we have tried to tell the story of the development of Real Analysis as completely as possible, our overriding goal was always to teach mathematics, not history. Thus we have necessarily left the history incomplete.
The interested student can fill in the gaps we have left by making use of the extensive resources that can be found at the MacTutor[2] history of mathematics repository.
All of the portraits of mathematicians used in this text have been taken from MacTutor.
MacTutor was created, and is maintained by Professor Edmund Robertson (Emeritus), and Professor John O'Connor (Emeritus), both of the University of St. Andrews in Scotland.

Desmos The interactive figures in Chapter 5 were created using the Desmos online calculator[3].

[2] mathshistory.st-andrews.ac.uk
[3] www.desmos.com

To the Instructor

The irony of this section is that it exists to tell you that this book was not written for you; it was written for your students. After all, we don't need to teach you about Real Analysis. You already understand the subject. The purpose of this text is to help your students make sense of the formal definitions, theorems, and proofs that they will encounter in your course. We do this by immersing the student in the story of how what is usually called Calculus evolved into modern Real Analysis. Our hope and intention is that this will help the student to appreciate why their intuitive understanding of topics encountered in calculus needs to be replaced by the formalism of Real Analysis.

The traditional approach to this topic (what we might call the "logical" story of Real Analysis), starts with a rigorous development of the real number system and uses this to provide rigorous definitions of limits, continuity, derivatives and integrals, and convergence of series; typically in that order. This is a perfectly legitimate story of Real Analysis and, logically, it makes the most sense. Indeed, this is a view of the subject that every mathematician-in-training should eventually attain. However, it is our contention that your students will appreciate the subject more, and hopefully retain it better, if they see how the subject developed from the intuitive notions of Leibniz, Newton and others in the seventeenth and eighteenth centuries to the more modern approach developed in the nineteenth and twentieth centuries. After all, they are coming at it from a viewpoint very similar to that of the mathematicians of the seventeenth and eighteenth centuries. Our goal is to bring them into the nineteenth and early twentieth centuries, mathematically speaking.

We hasten to add that this is not a history of analysis book. It is an introductory textbook on Real Analysis which uses the historical context of the subject to frame the concepts and to show why mathematicians felt the need to develop rigorous, non-intuitive definitions to replace their intuitive notions.

You will notice that most of the problems are embedded in the chapters, rather than lumped together at the end of each chapter. This is done to provide a context for the problems which, for the most part, are presented on an as-needed basis.

Thus the proofs of nearly all of the theorems appear as problems in the text. Of course, it would be very unfair to ask most students at this level to prove, say, the Bolzano-Weierstrass Theorem without some sort of guidance. So in each case we provide an outline of the proof and the subsequent problem will be to use the outline to develop a formal proof. Proof outlines will become less detailed as the students progress. We have found that this approach helps students develop their proof writing skills.

We state in the text, and we encourage you to emphasize to your students, that often they will use the results of problems as tools in subsequent problems.

Trained mathematicians do this naturally, but it is our experience that this is still somewhat foreign to students who are used to simply "getting the problem done and forgetting about it."

The problems range from the fairly straightforward to the more challenging. Some of them require the use of a computer algebra system (for example, to plot partial sums of a power series). These tend to occur earlier in the book where we encourage the students to use technology to explore the wonders of series. A number of these problems can be done on a sufficiently advanced graphing calculator or even on Wolfram Alpha, so you should assure your students that they do not need to be super programmers to do this. Of course, this is up to you.

A testing strategy we have used successfully is to assign more time consuming problems as collected homework and to assign other problems as possible test questions. Students could then be given some subset of these (verbatim) as an in-class test. Not only does this make test creation more straightforward, but it allows the opportunity to ask questions that could not reasonably be asked otherwise in a timed setting. Our experience is that this does not make the tests "too easy," and there are worse things than having students study by working together on possible test questions beforehand. If you are shocked by the idea of giving students all of the possible test questions ahead of time, think of how much (re)learning you did studying the list of possible questions you *knew* would be asked on a qualifying exam.

In the end, use this book as you see fit. We believe your students will find it readable, as it is intended to be, and we are confident that it will help them to make sense out of the rigorous, non-intuitive definitions and theorems of Real Analysis and help them to develop their proof-writing skills.

If you have suggestions for improvement, comments or criticisms of our text please contact us at the email addresses below. We appreciate any feedback you can give us on this.

Thank you.

Robert R. Rogers
robert.rogers@fredonia.edu

Eugene C. Boman
ecb5@psu.edu

Contents

Back Matter

Part I

In Which We Raise A Number Of Questions

Chapter 1

Prologue: Three Lessons Before We Begin

1.1 Lesson One

Get a pad of paper and write down the answer to this question: What is . . . No, really. We're serious. *Get a writing pad.* We'll wait.

Comment. We *really* are serious about this. Get a pad of paper!

Got it? Good. Now write down your answer to this question: What is a number? Don't think about it. Don't analyze it. Don't consider it. Just write down the best answer you can without thinking. You are the only person who ever needs to see what you've written.

Done? Good.

Now consider this: All of the objects listed below are "numbers" in a sense we will not make explicit here. How many of them does your definition include?

- 1
- -1
- 0
- $3/5$
- $\sqrt{2}$
- $\sqrt{-1}$
- i^i
- e^{5i}
- $4 + 3i - 2j + 6k$ (this is called a quaternion)
- dx (this is the differential you learned all about in calculus)
- $\begin{pmatrix} 1 & 2 \\ -2 & 1 \end{pmatrix}$ (yes, matrices can be considered numbers).

Surely you included 1. Almost surely you included $3/5$. But what about 0? -1? Does your definition include $\sqrt{2}$? Do you consider dx a number? Leibniz did. Any of the others? (And, yes, they really are all "numbers.")

The lesson in this little demonstration is this: You don't really have a clear notion of what we mean when we use the word "number." And this is fine. Not knowing is acceptable.

Comment. Sometimes it is even encouraged.

A principal goal of this course of study is to rectify this, at least a little bit. When the course is over you may or may not be able to give a better definition of the word "number" but you will have a deeper understanding of the real numbers at least. That is enough for now.

1.2 Lesson Two

Read and understand the following development of the Quadratic Formula.

Suppose $a \neq 0$. If

$$ax^2 + bx + c = 0 \tag{I.1}$$

then

$$x^2 + \frac{b}{a}x = -\frac{c}{a} \tag{I.2}$$

Now let $x = y - \frac{b}{2a}$ giving

$$y^2 = -\frac{c}{a} + \frac{b^2}{4a^2} \tag{I.3}$$

$$y = \frac{\pm\sqrt{b^2 - 4ac}}{2a} \tag{I.4}$$

$$x = \frac{-b \pm \sqrt{b^2 - 4ac}}{2a} \tag{I.5}$$

Were you able to follow the argument? Probably the step from equation (I.1) to equation (I.2) presented no difficulties. But what about the next step? Do you see where equation (I.3) came from? If so, good for you. Most students, in fact most mathematicians, cannot make that step in their heads. But are you sure? Is there, perhaps, some small detail you've overlooked?

Check to see.

That is, let $x = y - \frac{b}{2a}$ in equation (I.2) and see if you can get equation (I.3). Do it on that handy pad of paper we told you to get out earlier. Do it now. We'll wait.

Comment. If you *still* haven't gotten out a pad of paper, give up now. You're going to fail this course. Seriously. Do you think we would spend so much time on this, that we would repeat it so many times, if it weren't important. GET OUT A PAD OF PAPER NOW! Last chance. You've been warned.

Done? Good.

Perhaps you haven't been able to fill in the details on your own. That's ok. Many people can't. If not then get help: from a classmate, a friend, your instructor, whomever. Unfortunately most people won't get help in this situation. Instead they will perceive this as "failure," hide it and berate themselves or the problem as "stupid." In short they will let their personal insecurities and demons overwhelm them. *Don't do this. Get help.* You are neither dumb nor incapable. There are a thousand reasons that on any given day you might not be able to solve this problem. But don't let a bad day interfere with the education you are here for. Get someone to help you over this hump. Later you will be able to help your helper in the same way. Really.

At this point we assume that you've successfully negotiated the transition from equation (I.2) to equation (I.5).

See? It really wasn't that bad after all. Just a lot of elementary algebra. Now that you've done it (or seen it done), it is easy to see that there really wasn't much there.

But this is the point! We left those computations out precisely because we knew that they were routine and that you could fill in the details. Moreover, filling in the details yourself gives you a little better insight into the computations. If we'd filled them in for you we would have robbed you of that insight. And we would have made this book longer than it needs to be. We don't want to do either of those things. If we fill in all of the details of every computation for you, you won't learn to have confidence in your ability to do them yourself. And this book will easily double in length.

So the lesson here is this: Keep that pad of paper handy whenever you are reading this (or any other) mathematics text. You will need it. Routine computations will often be skipped. But calling them "routine" and skipping them does not mean that they are unimportant. If they were truly unimportant we would leave them out entirely.

Moreover "routine" does not mean "obvious." Every step we took in the development of the Quadratic Formula was "routine." But even routine computations need to be understood and the best way to understand them is to do them. This is the way to learn mathematics; it is the *only* way that really works. Don't deprive yourself of your mathematics education by skipping the most important parts.

Comment. If you didn't fill in those details then you are depriving yourself of the education you are here to obtain. This is sad. There is a good reason for putting these three lessons first. Stop wasting your time and intellect! Go do it now.

As you saw when you filled in the details of our development of the Quadratic Formula the substitution $x = y - \frac{b}{2a}$ was crucial because it turned

$$x^2 + \frac{b}{a}x + \frac{c}{a} = 0$$

into

$$y^2 = k$$

where k depends only on a, b, and c. In the sixteenth century a similar technique was used by Ludovico Ferrari (1522-1565) to reduce the general cubic equation

$$ax^3 + bx^2 + cx + d = 0 \tag{I.6}$$

into the so-called "depressed cubic"

$$y^3 + py + q = 0$$

where p, and q depend only on a, b, c, and d.

The general depressed cubic had previously been solved by Tartaglia (the Stutterer, 1500-1557) so converting the general cubic into a depressed cubic provided a path for Ferrari to compute the "Cubic Formula" — like the Quadratic Formula but better.

Comment. It is not entirely clear why eliminating the quadratic term should be depressing, but there it is.

Figure 1.2.1 Tartaglia[4], "The Stutterer"

Ferrari also knew how to compute the general solution of the "depressed quartic" so when he and his teacher Girolomo Cardano (1501-1576) figured out how to depress a general quartic they had a complete solution of the general quartic as well.

Figure 1.2.2 Girolomo Cardano[5]

Alas, their methods broke down entirely when they tried to solve the general quintic equation. Unfortunately the rest of this story belongs in a course on Abstract Algebra, not Real Analysis. But the lesson in this story applies to all of mathematics: Every problem solved is a new theorem which then becomes a tool for later use. Depressing a cubic would have been utterly useless had not Tartaglia had a solution of the depressed cubic in hand. The technique they used, with slight modifications, then allowed for a solution of the general quartic as well.

Keep this in mind as you proceed through this course and your mathematical education. Every problem you solve is really a theorem, a potential tool that you can use later. We have chosen the problems in this text deliberately with this in mind. Don't just solve the problems and move on. Just because you have solved a problem does not mean you should stop thinking about it. Keep thinking about the problems you've solved. Internalize them. Make the ideas your own so that when you need them later you will have them at hand to use.

Problem 1.2.3

a Find M so that the substitution $x = y - M$ depresses equation (I.6), the general cubic equation. Then find p and q in terms of a, b, c, and d.

b Find K so that the substitution $x = y - K$ depresses the general quartic equation. Make sure you demonstrate how you obtained that value or

[4] mathshistory.st-andrews.ac.uk/Biographies/Tartaglia/

[5] mathshistory.st-andrews.ac.uk/Biographies/Cardan

why it works (if you guessed it).

c Find N so that the substitution $x = y - N$ depresses a polynomial of degree n. Ditto on showing that this value works or showing how you obtained it.

□

Problem 1.2.4 Another Derivation of the Quadratic Formula. Here is yet another way to solve a quadratic equation. Read the development below with pencil and paper handy. Confirm all of the computations that are not completely transparent to you. Then use your notes to present the solution with *all* steps filled in.

Comment. Be sure you are clear on the purpose of this problem before you begin. This is not about solving the Quadratic Equation. You already know how to do that. Our purpose here is to give you practice filling in the skipped details of mathematical exposition. We've chosen this particular problem because it should be a comfortable setting for you, but this particular solution is probably outside of your previous experience.

Suppose that r_1 and r_2 are solutions of $ax^2 + bx + c = 0$. Without loss of generality suppose that $a > 0$. Suppose further that $r_1 \geq r_2$. Then

$$\begin{aligned} ax^2 + bx + c &= a(x - r_1)(x - r_2) \\ &= a\left[x^2 - (r_1 + r_2)x + (r_1 + r_2)^2 - (r_1 - r_2)^2 - 3r_1 r_2\right]. \end{aligned}$$

Therefore

$$r_1 + r_2 = -\frac{b}{a} \tag{I.7}$$

and

$$r_1 - r_2 = \sqrt{\left(\frac{b}{a}\right)^2 - \frac{4c}{a}}. \tag{I.8}$$

Equations (I.7) and (I.8) can be solved simultaneously to yield

$$\begin{aligned} r_1 &= \frac{-b + \sqrt{b^2 - 4ac}}{2a} \\ r_2 &= \frac{-b - \sqrt{b^2 - 4ac}}{2a}. \end{aligned}$$

□

1.3 Lesson Three

In the hustle and bustle of a typical college semester, with a lot of demands on your time and very little time to think, it becomes very easy to see each problem you solve as a small, isolated victory and then move on to the next challenge. This is understandable. Each problem you solve *is* a small victory and you've every right to be proud of it. But it is not isolated and it is a mistake to think that it is.

Figure 1.3.1 George Polya[6]

In his book *How to Solve It* the mathematician and teacher George Polya gave four steps for problem solving. The steps may be paraphrased as

1. Understand the problem.
2. Formulate a plan.
3. Execute the plan.
4. Reflect on what you've done.

This process is iterative. That is, once a plan is formulated and executed we often find that our plan was not up to the task. So we have to ask what went wrong, form a new plan and try again. This is the fourth step: Reflect on what you've done.

Almost everyone remembers this fourth step when their plan *doesn't* work. After all, you've got to try again so you have to ask what went wrong. But it is all too easy to neglect that crucial fourth step when the plan succeeds. In that case, flush with success we usually move on to the next problem and start over from scratch.

This is a mistake. Having solved a problem is no reason to stop thinking about it.

That fourth step is at least as important when we have succeeded as when we have failed. Each time you solve a problem stop and ask yourself a few questions:

- Are there any easy consequences that follow from the result?
- How does it fit into the broader scheme of other problems you have solved?
- How might it be used in the future?

Also, notice the structure of the problem. Some assumptions had to be made. What were they? Were they all necessary? That is, did your solution use everything that was assumed? If not, you may have something considerably more general than it at first appears. What is that more general statement? Even if you used all of the assumptions, was that really necessary? Can you solve a similar problem with weaker assumptions?

Take a moment to pack all of these questions (and their answers) away in your mind so that when you see something similar in the future you will be reminded of it. *Don't* solve any problem and then forget it and move on. The nature of mathematics is cumulative. Remember, you are not here to

[6] mathshistory.st-andrews.ac.uk/Biographies/Polya/

accumulate grade points. You are here to learn and understand the concepts and methods of mathematics, to gain "mathematical maturity." Part of that maturation process is the accumulation of a body of facts (theorems), and techniques that can be used to prove new theorems (solve new problems).

This text has been written with the maturation process in mind. You will frequently find that the problems you solve today can be used to good effect in the ones you attempt tomorrow, but only if you remember them. So take a moment after you've solved each problem to think about how it fits into the patterns you already know. This is important enough to bear repeating: *A problem, once solved, becomes a tool for solving subsequent problems!*

The purpose of the following sequence of problems is to help you become accustomed to this notion (if you aren't already). It is a progression of results about prime numbers. As you probably recall, a prime number is any integer greater than 1 whose only factors are itself and 1. For example, 2, 3, 5, 7, 11 are prime, while 4, 6, 9 are not. A major result about prime numbers is the following:

Theorem 1.3.2 ***The Fundamental Theorem of Arithmetic***

Any integer greater than 1 *is either prime or it is a product of prime numbers. Furthermore, this prime decomposition is unique up to the order of the factors.*

We will not prove this, but we will use it as a starting point to examine the following problems. As you do these problems, notice how subsequent problems make use of the previous results.

Notice that the notation $p\,|a$ simply means that the integer p divides the integer a with no remainder.

Problem 1.3.3 Fermat's Little Theorem, step 1. Let p be a prime number and a, b positive integers such that $p\,|(a \cdot b)$. Show that $p\,|a$ or $p\,|b$.

Hint.

If $p\,|a$ then we are done. If not then notice that p is a prime factor of $a \cdot b$. What does the Fundamental Theorem of Arithmetic say about the prime factors of $a \cdot b$ compared to the prime factors of a and b? □

Problem 1.3.4 Fermat's Little Theorem, step 2. Let p be a prime number and let $a_1, a_2, \ldots, a_n$ be positive integers such that $p\,|\,(a_1 \cdot a_2 \cdot a_3 \cdot \ldots \cdot a_n)$. Show that $p\,|a_k$ for some $k \in \{1, 2, 3, \ldots, n\}$.

Hint.

Use induction on n and the result of the previous problem. □

Problem 1.3.5 Fermat's Little Theorem, step 3. Let p be a prime number and let k be an integer with $1 \leq k \leq p-1$. Prove that $p\,\big|\binom{p}{k}$, where $\binom{p}{k}$ is the binomial coefficient $\frac{p!}{k!(p-k)!}$.

Hint.

We know $p\,|p!$, so $p\,|\binom{p}{k}k!(p-k)!$. How does the previous result apply? □

We now have all the machinery in place to prove one of the really cool theorems from number theory.

Theorem 1.3.6 ***Fermat's Little Theorem***

Let p be any prime number. Then $p\,|(n^p - n)$ for all positive integers n.

Problem 1.3.7 Fermat's Little Theorem. Prove Fermat's Little Theorem.

Hint.

Use induction on n. To get from n to $n+1$, use the binomial theorem on $(n+1)^p$. □

Fermat's Little Theorem is the foundational basis for a number of results in number theory and encryption.

Chapter 2

Numbers, Real ($\mathbb{R}$) and Rational ($\mathbb{Q}$)

The set of real numbers (denoted, $\mathbb{R}$) is badly named. The real numbers are no more or less real — in the non-mathematical sense that they exist — than any other set of numbers, just like the set of rational numbers ($\mathbb{Q}$), the set of integers ($\mathbb{Z}$), or the set of natural numbers ($\mathbb{N}$). The name "real numbers" is (almost) an historical anomaly not unlike the name "Pythagorean Theorem" which was actually known and understood long before Pythagoras lived.

When calculus was being invented in the 17th century, numbers were thoroughly understood, or so it was believed.

Comment. Some would say "re-invented." See [13], or [9].

They were, after all, just numbers. Combine them. We call that addition. If you add them repeatedly we call it multiplication. Subtraction and division were similarly understood.

It was (and still is) useful to visualize these things in a more concrete way. If we take a stick of length 2 and another of length 3 and lay them end-to-end we get a length of 5. This is addition. If we lay them end-to-end but at right angles then our two sticks are the length and width of a rectangle whose area is 6. This is multiplication.

Of course measuring lengths with whole numbers has limitations, but these are not hard to fix. If we have a length (stick) of length 1 and another of length 2, then we can find another whose length when compared to 1 is the same (has the same proportion) as 1 is to 2. That number of course, is 1/2.

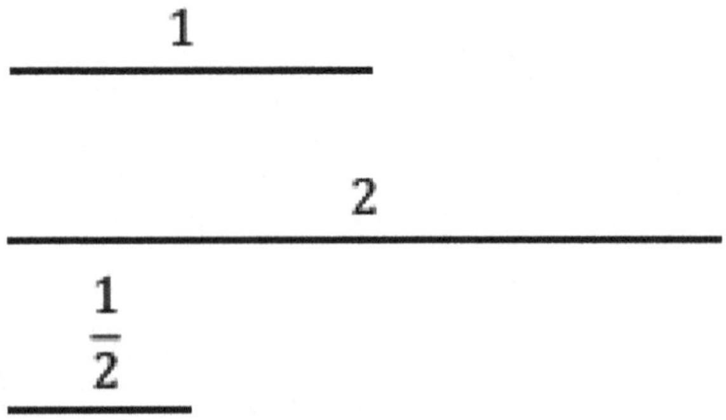

Figure 2.0.1

Notice how fraction notation reflects the operation of comparing 1 to 2. This comparison is usually referred to as the *ratio* of 1 to 2 so numbers of this sort are called *rational numbers.* The set of rational numbers is denoted $\mathbb{Q}$ for quotients. In grade school they were introduced to you as fractions. Once fractions are understood, this visualization using line segments (sticks) leads quite naturally to their representation with the rational number line.

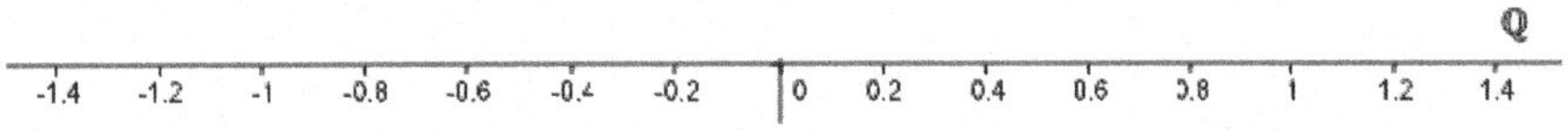

Figure 2.0.2 The Rational Number Line

This seems to work as a visualization because the rational numbers and the points on a line seem to share certain properties. Chief among these is that between any two points on the rational line there is another point, just as between any two rational numbers there is another rational number.

Problem 2.0.3 Let $a, b, c, d \in \mathbb{N}$ and find a rational number between a/b and c/d. □

This is all very clean and satisfying until we examine it just a bit closer. Then it becomes quite mysterious. Consider again the rational numbers a/b and c/d. If we think of these as lengths we can ask, "Is there a third length, say α, such that we can divide a/b into M pieces, each of length α *and also* divide c/d into N pieces each of length α?" A few minutes thought should convince you that this is the same as the problem of finding a common denominator so $\alpha = \frac{1}{bd}$ will work nicely. (Confirm this yourself.)

You may be wondering what we're making all of this fuss about. *Obviously* this is *always* true. In fact the previous paragraph gives an outline of a very nice little proof of this. Here are the theorem and its proof presented formally.

Theorem 2.0.4 *Suppose a, b, c, and d are integers. There is a number $\alpha \in \mathbb{Q}$ such that $M\alpha = a/b$ and $N\alpha = c/d$ where M and N are also integers.*

Proof. To prove this theorem we will display α, M and N. It is your responsibility to confirm that these actually work. Here they are: $\alpha = 1/bd$, $M = ad$, and $N = cb$. ■

Problem 2.0.5 Confirm that α, M, and N as given in the proof of Theorem 2.0.4 satisfy the requirements of the theorem. □

Theorem 2.0.4 suggests the following very deep and important question: Are there lengths which can *not* be expressed as the ratio of two integer lengths? The answer, of course, is yes. Otherwise we wouldn't have asked the question.

One of the best known examples of such a number is the circumference of a circle with diameter 1. This is the number usually denoted by π. But circles are extremely complex objects — they only seem simple because they are so familiar. Arising as it does from a circle, you would expect the number π to be very complex as well and this is true. In fact π is an exceptionally weird number for a variety of reasons. Let's start with something a little easier to think about.

Squares are simple. Two sets of parallel lines at right angles, all of the same length. What could be simpler? If we construct a square with sides having length 1 then its diagonal has length $\sqrt{2}$.

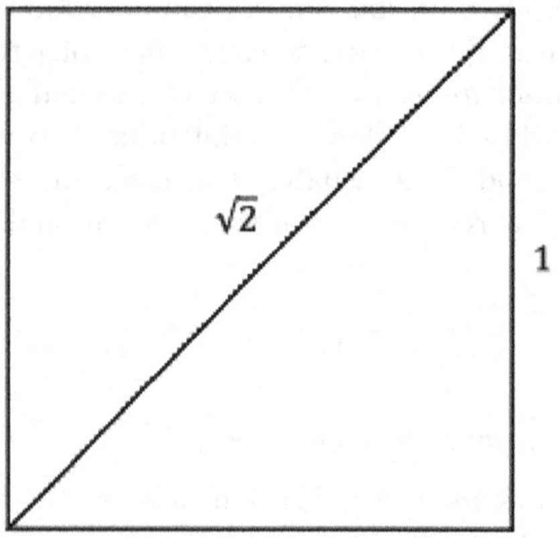

Figure 2.0.6 A construction of $\sqrt{2}$

This is a number which cannot be expressed as the *ratio* of two integers. That is, it is *irrational.* This has been known since ancient times, but it is still quite disconcerting when first encountered. It seems so counter-intuitive that the intellect rebels. "This can't be right," it says. "That's just crazy!"

Nevertheless it is true and we can prove it is true as follows.

What happens if we suppose that the square root of two *can* be expressed as a ratio of integers? We will show that this leads irrevocably to a conclusion that is manifestly not true.

Suppose $\sqrt{2} = a/b$ where a and b are integers. Suppose further that the fraction a/b is in lowest terms. *This assumption is crucial because if a/b is in lowest terms we know that at most only one of them is even.*

So

$$\frac{a}{b} = \sqrt{2}.$$

Squaring both sides gives:

$$a^2 = 2b^2.$$

Therefore a^2 is even. But if a^2 is even then a must be even also (why?). If a is even then $a = 2k$ for some integer k. Therefore

$$4k^2 = 2b^2 \text{ or}$$
$$2k^2 = b^2.$$

Therefore b^2 is also even and so b must be even too. But this is impossible. We've just concluded that a and b are both even and this conclusion follows directly from our initial assumption that at most one of them could be even.

This is nonsense. Where is our error? It is not in any single step of our reasoning. That was all solid. Check it again to be sure.

Therefore our error must be in the initial assumption that $\sqrt{2}$ could be expressed as a fraction. That assumption must therefore be false. In other words, $\sqrt{2}$ cannot be so expressed.

Problem 2.0.7 Irrational Numbers. Show that each of the following numbers is irrational:

(a) $\sqrt{3}$

(b) $\sqrt{5}$

(c) $\sqrt[3]{2}$

(d) i $(= \sqrt{-1})$

(e) The square root of every positive integer which is not the square of an integer.

□

The fact that $\sqrt{2}$ is not rational is cute and interesting, but unless, like the Pythagoreans of ancient Greece, you have a strongly held religious conviction that all numbers are rational, it does not seem terribly important. On the other hand, the very existence of $\sqrt{2}$ raises some interesting questions. For example what can the symbol $4^{\sqrt{2}}$ possibly mean? If the exponent were a rational number, say m/n, then clearly $4^{m/n} = \sqrt[n]{4^m}$. But since $\sqrt{2} \neq m/n$ for *any* integers m and n how do we interpret $4^{\sqrt{2}}$? Does it have any meaning at all? The more you think about this, the more puzzling the existence of irrational numbers becomes. Suppose for example we reconsider the construction of a line segment of length $\sqrt{2}$. It is clear that the construction works and that we really can build such a line segment. It exists.

Repeat the construction but this time let's put the base side on the rational line.

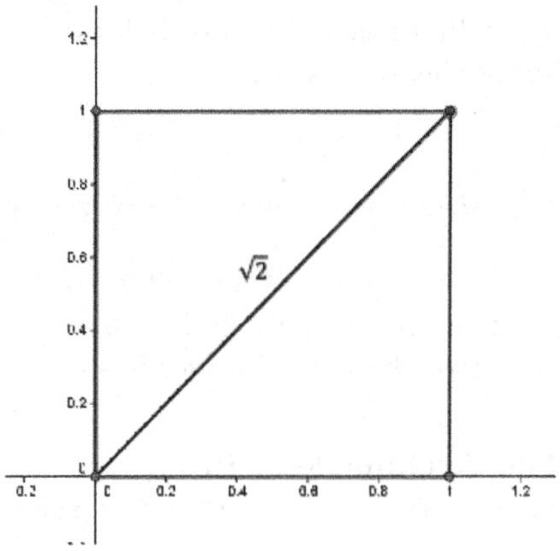

Figure 2.0.8

We know that the diagonal of this square is $\sqrt{2}$ as indicated. And we know that $\sqrt{2}$ is not a rational number.

Now leave the diagonal pinned at $(0, 0)$ but allow it to rotate down so that it coincides with the $x-$axis.

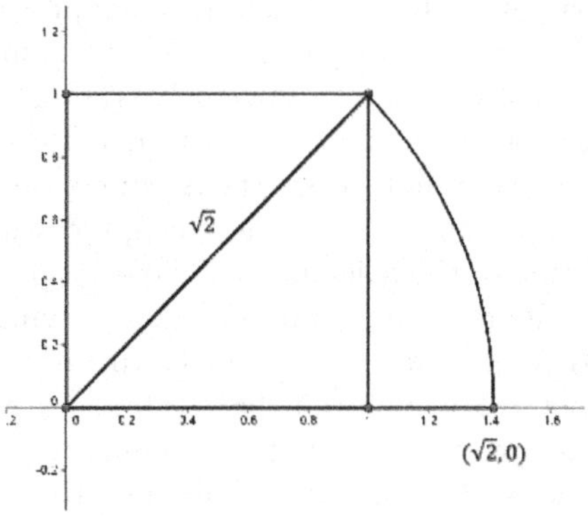

Figure 2.0.9

The end of our diagonal will trace out an arc of the circle with radius $\sqrt{2}$. When the diagonal coincides with the $x-$axis, its endpoint will obviously be the point $(\sqrt{2}, 0)$ as shown.

But wait! We're using the *rational* number line for our $x-$axis. That means the only points on the $x-$axis are those that correspond to rational numbers (fractions). But we know that $\sqrt{2}$ is not rational! Conclusion: There is no point $(\sqrt{2}, 0)$. It simply doesn't exist.

Put differently, there is a hole in the rational number line right where $\sqrt{2}$ should be.

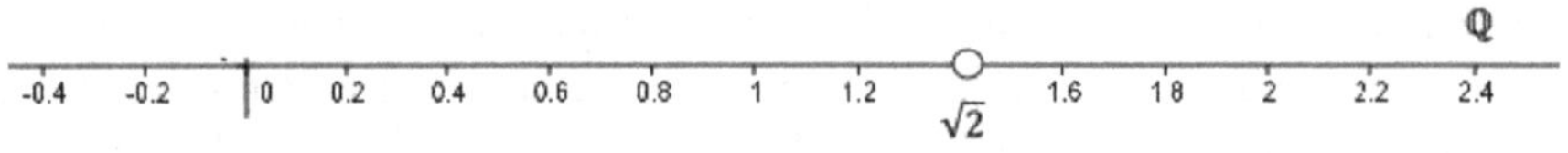

Figure 2.0.10

This is weird!

Recall that between any two rational numbers there is always another. This fact is what led us to represent the rational numbers with a line in the first place.

But it's even worse than that. It's straightforward to show that $\sqrt{3}$, $\sqrt{5}$, etc. are all irrational too. So are π and e, though they aren't as easy to show. It seems that the rational line has a bunch of holes in it. Infinitely many.

And yet, the following theorem is true

Theorem 2.0.11

a *Between any two distinct real numbers there is a rational number.*

b *Between any two distinct real numbers there is an irrational number.*

Both parts of this theorem rely on a judicious use of what is now called the Archimedean Property of the Real Number System, which can be formally stated as follows.

Principle 2.0.12 The Archimedean Property. *Given any two positive real numbers, a and b, there is a positive integer, n such that $na > b$.*

Physically this says that we can empty an ocean b with a teaspoon a, provided we are willing to use the teaspoon a large number of times n.

This is such an intuitively straightforward concept that it is easy to accept it without proof. Until the invention of calculus, and even for some time after that, it was simply assumed. However as the foundational problems posed by the concepts of calculus were understood and solved we were eventually led to a deeper understanding of the complexities of the real number system. The Archimedean Property is no longer taken as an unproved axiom, but rather it is now understood to be a consequence of other axioms. We will show this later, but for now we will accept it as obviously true just as Archimedes did.

With the invention of calculus, mathematicians of the seventeenth century began to use objects which didn't satisfy the Archimedean Property (in fact, so did Archimedes). As we shall see in the next chapter, when Leibniz wrote the first paper on his version of the calculus, he followed this practice by explicitly laying out rules for manipulating infinitely small quantities (infinitesimals). These were taken to be actual numbers which are not zero and yet smaller than any real number. The notation he used was $\mathrm{d}x$ (an infinitely small displacement in the x direction), and $\mathrm{d}y$ (an infinitely small displacement in the y direction). These symbols should look familiar to you. They are the same $\mathrm{d}y$ and $\mathrm{d}x$ used to form the derivative symbol $\frac{\mathrm{d}y}{\mathrm{d}x}$ that you learned about in calculus.

Mathematicians of the seventeenth and eighteenth centuries made amazing scientific and mathematical progress exploiting these infinitesimals, even though they were foundationally suspect. No matter how many times you add the infinitesimal dx to itself the result will not be bigger than, say 10^{-1000}, which is very bizarre.

When foundational issues came to the forefront, infinitesimals fell somewhat out of favor. You probably didn't use them very much in calculus. Most of the time you probably used the prime notation, $f'(x)$ introduced by Lagrange in the eighteenth century. Some of the themes in this book are: Why differentials fell out of favor, what were they replaced with and how the modern notations you learned in calculus evolved over time.

To conclude this aside on the Archimedean Property, the idea of infinitesimals was revisited in the twentieth century by the logician Abraham Robinson in [12]. Robinson was able to put the idea of infinitesimals on a solid logical foundation. But in the 18th century, the existence of infinitesimal numbers was shaky to say the very least. However this did not prevent mathematicians from successfully exploiting these infinitely small quantities.

We will come back to this saga in later chapters, but for now we return to Theorem 2.0.11.

Sketch of Proof. We will outline the proof of part (a) of Theorem 2.0.11 and indicate how it can be used to prove part b.
Let α and β be real numbers with $\alpha > \beta$. There are two cases.

- *Case 1:.*

  $\alpha - \beta > 1$. In this case there is at least one integer between α and β. Since integers are rational we are done.

- *Case 2:.*

  $\alpha - \beta \leq 1$. In this case, by the Archimedean Property there is a positive integer, say n, such that $n(\alpha - \beta) = n\alpha - n\beta > 1$. Now there will be an integer between $n\alpha$ and $n\beta$. You should now be able to find a rational number between α and β.

■

For part b, divide α and β by any positive irrational number and apply part a. There are a couple of details to keep in mind. These are considered in the following problem.

Problem 2.0.13

(a) Prove that the product of a nonzero rational number and an irrational number is irrational.

(b) Use the result of part (a) to prove Theorem 2.0.11.

□

As a practical matter, the existence of irrational numbers isn't really very important. In light of Theorem 2.0.11, any irrational number can be approximated arbitrarily closely by a rational number. So if we're designing a bridge and $\sqrt{2}$ is needed we just use 1.414 instead. The error introduced is less than $0.001 = 1/1000$ so it probably doesn't matter.

But from a theoretical point of view this is devastating. When calculus was invented, the rational numbers were suddenly not up to the task of justifying the concepts and operations we needed to work with.

Newton explicitly founded his version of calculus on the assumption that we can think of variable quantities as being generated by a continuous motion.

If our number system has holes in it such continuous motion is impossible because we have no way to jump over the gaps. So Newton simply postulated that there were no holes. He filled in the hole where $\sqrt{2}$ should be. He simply said, yes there is a number there called $\sqrt{2}$ and he did the same with all of the other holes.

To be sure there is no record of Newton explicitly saying, "Here's how I'm going to fill in the holes in the rational number line." Along with everyone else at the time, he simply assumed there were no holes and moved on. It took about 200 years of puzzling and arguing over the contradictions, anomalies and paradoxes to work out the consequences of that apparently simple assumption. The task may not yet be fully accomplished, but by the 20th century the properties of the real number system (ℝ) as an extension of the rational number system (ℚ) were well understood. Here are both systems visualized as lines:

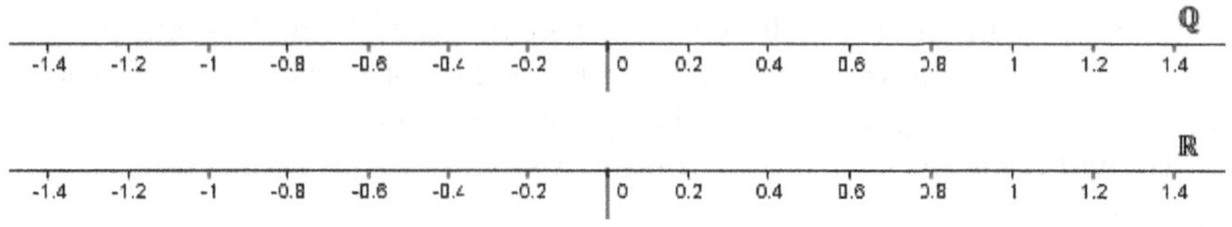

Figure 2.0.14 ℝ and ℚ

Impressive, no?

The reason they look alike, except for the labels ℝ and ℚ of course, is that our ability to draw sketches of the objects we're studying utterly fails when we try to sketch ℝ, as different from ℚ. All of the holes in ℚ really are there, but the non-holes are packed together so closely that we can't separate them in a drawing. This inability to sketch the objects we study will be a frequent source of frustration.

Of course, this will not stop us from drawing sketches. When we do, our imaginations will save us because it *is* possible to *imagine* ℚ as distinct from ℝ. But put away the idea that a sketch is an accurate representation of anything. At best our sketches will only be aids to the imagination.

So, at this point we will simply assume the existence of the real numbers. We will assume also that they have all of the properties that we are used to. This is perfectly acceptable as long as we make our assumptions explicit. However we need to be aware that, so far, the existence and properties of the real numbers is an *assumption* that has not been logically derived. Any time we make an assumption we need to be prepared to either abandon it completely if we find that it leads to nonsensical results, or to re-examine the assumption in the light of these results to see if we can find another assumption that subsumes the first and explains the (apparently) nonsensical results.

2.1 Additional Problems

Problem 2.1.1 Determine if each of the following is always rational or always irrational. Justify your answers.

(a) The sum of two rational numbers.

(b) The sum of two irrational numbers.

(c) The sum of a rational and an irrational number.

□

Problem 2.1.2 Is it possible to have two rational numbers, a and b, such that a^b is irrational? If so, display an example of such a and b. If not, prove that it is not possible. □

Problem 2.1.3 Decide if it is possible to have two irrational numbers, a and b, such that a^b is rational. Prove it in either case. □

Chapter 3

Calculus in the 17th and 18th Centuries

3.1 Newton and Leibniz Get Started

3.1.1 Leibniz's Calculus Rules

Figure 3.1.1 Gottfried Wilhelm Leibniz[7]

The rules for calculus were first laid out in Gottfried Wilhelm Leibniz's 1684 paper *Nova methodus pro maximis et minimis, itemque tangentibus, quae nec fractas nec irrationales, quantitates moratur, et singulare pro illi calculi genus* (A New Method for Maxima and Minima as Well as Tangents, Which is Impeded Neither by Fractional Nor by Irrational Quantities, and a Remarkable Type of Calculus for This). Leibniz started with subtraction. That is, if x_1 and x_2 are very close together then their difference, $\Delta x = x_2 - x_1$, is very small. He expanded this idea to say that if x_1 and x_2 are *infinitely* close together (but still distinct) then their difference, $\mathrm{d}x$, is infinitesimally small (but not zero).

Calculus Differentialis. This translates, loosely, as the calculus of differences.

This idea is logically very suspect and Leibniz knew it. But he also knew that when he used his *calculus differentialis* he was getting correct answers to some very hard problems. So he persevered.

Leibniz called both Δx and $\mathrm{d}x$ "differentials" (Latin for difference) because he thought of them as, essentially, the same thing. Over time it has become

[7] mathshistory.st-andrews.ac.uk/Biographies/Leibniz/

customary to refer to the infinitesimal $\mathrm{d}x$ as a differential, reserving "difference" for the finite case, Δx. This is why calculus is often called "differential calculus."

In his paper Leibniz gave rules for dealing with these infinitely small differentials. Specifically, given a variable quantity x, dx represented an infinitesimal change in x. Differentials are related via the slope of the tangent line to a curve. That is, if $y = f(x)$, then $\mathrm{d}y$ and $\mathrm{d}x$ are related by

$$\mathrm{d}y = \text{ (slope of the tangent line) } \cdot \mathrm{d}x.$$

Leibniz then divided by $\mathrm{d}x$ giving

$$\frac{\mathrm{d}y}{\mathrm{d}x} = \text{ (slope of the tangent line)}.$$

The elegant and expressive notation Leibniz invented was so useful that it has been retained through the years despite some profound changes in the underlying concepts. For example, Leibniz and his contemporaries would have viewed the symbol $\frac{\mathrm{d}y}{\mathrm{d}x}$ as an actual quotient of infinitesimals, whereas today we define it via the limit concept first suggested by Newton.

As a result the rules governing these differentials are very modern in appearance:

$$\mathrm{d}(\text{ constant }) = 0$$
$$\mathrm{d}(z - y + w + x) = \mathrm{d}z - \mathrm{d}y + \mathrm{d}w + \mathrm{d}x$$
$$\mathrm{d}(xv) = x\,\mathrm{d}v + v\,\mathrm{d}x$$
$$\mathrm{d}\left(\frac{v}{y}\right) = \frac{y\,\mathrm{d}v - v\,\mathrm{d}y}{yy}$$

and, when a is an integer:

$$\mathrm{d}(x^a) = ax^{a-1}\,\mathrm{d}x.$$

Leibniz states these rules without proof: ". . . the demonstration of all this will be easy to one who is experienced in such matters . . .." As an example, mathematicians in Leibniz's day would be expected to understand intuitively that if c is a constant, then $d(c) = c - c = 0$. Likewise, $d(x + y) = dx + dy$ is really an extension of $(x_2 + y_2) - (x_1 + y_1) = (x_2 - x_1) + (y_2 - y_1)$.

3.1.2 Leibniz's Approach to the Product Rule

The explanation of the product rule using differentials is a bit more involved, but Leibniz expected that mathematicans would be fluent enough to derive it. The product $p = xv$ can be thought of as the area of the following rectangle

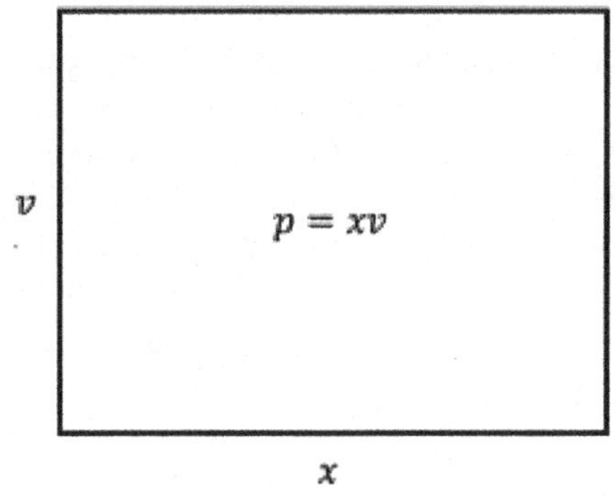

Figure 3.1.2

With this in mind, $\mathrm{d}p = \mathrm{d}(xv)$ can be thought of as the change in area when x is changed by $\mathrm{d}x$ and v is changed by $\mathrm{d}v$. This can be seen as the L shaped region in the following drawing.

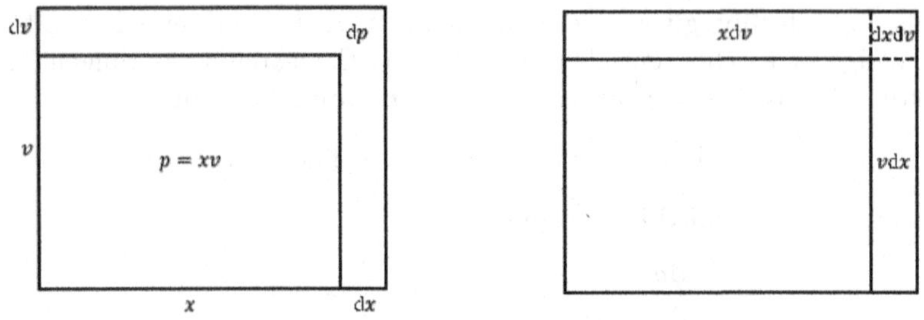

Figure 3.1.3

By dividing the L shaped region into 3 rectangles we obtain

$$\mathrm{d}(xv) = x\,\mathrm{d}v + v\,\mathrm{d}x + \mathrm{d}x\ \mathrm{d}v. \tag{I.1}$$

Even though $\mathrm{d}x$ and $\mathrm{d}v$ are infinitely small, Leibniz reasoned that $\mathrm{d}x\ \mathrm{d}v$ is *even more* infinitely small (quadratically infinitely small?) compared to $x\,\mathrm{d}v$ and $v\,\mathrm{d}x$ and can thus be ignored leaving

$$\mathrm{d}(xv) = x\,\mathrm{d}v + v\,\mathrm{d}x.$$

You should feel some discomfort at the idea of simply tossing the product $\mathrm{d}x\ \mathrm{d}v$ aside because it is "comparatively small." This means you have been well trained, and have thoroughly internalized Newton's dictum [10]: "The smallest errors may not, in mathematical matters, be scorned." It is logically untenable to toss aside an expression just because it is small. Even less so should we be willing to ignore an expression on the grounds that it is "infinitely smaller" than another quantity which is itself "infinitely small."

Newton and Leibniz both knew this as well as we do. But they also knew that their methods *worked*. They gave verifiably correct answers to problems which had, heretofore, been completely intractable. It is the mark of their genius that both men persevered in spite of the very evident difficulties their methods entailed.

3.1.3 Newton's Approach to the Product Rule

In the Principia, Newton "proved" the Product Rule as follows: Let x and v be "flowing quantites" and consider the rectangle, R, whose sides are x and v. R is also a flowing quantity and we wish to find its fluxion (derivative) at any time.

Newton's 'Method of Fluxions'. Newton's approach to calculus — his 'Method of Fluxions' — depended fundamentally on motion. That is, he viewed his variables (fluents) as changing (flowing or fluxing) in time. The rate of change of a fluent he called a fluxion. As a foundation both Leibniz's and Newton's approaches have fallen out of favor, although both are still universally used as a conceptual approach, a "way of thinking," about the ideas of calculus.

First increment x and v by $\frac{\Delta x}{2}$ and $\frac{\Delta v}{2}$ respectively. Then the corresponding increment of R is

$$\left(x + \frac{\Delta x}{2}\right)\left(v + \frac{\Delta v}{2}\right) = xv + x\frac{\Delta v}{2} + v\frac{\Delta x}{2} + \frac{\Delta x \Delta v}{4}. \tag{I.2}$$

Now decrement x and v by the same amounts:

$$\left(x - \frac{\Delta x}{2}\right)\left(v - \frac{\Delta v}{2}\right) = xv - x\frac{\Delta v}{2} - v\frac{\Delta x}{2} + \frac{\Delta x \Delta v}{4}. \tag{I.3}$$

Subtracting the right side of equation (I.3) from the right side of equation (I.2) gives

$$\Delta R = x\Delta v + v\Delta x$$

which is the total change of $R = xv$ over the intervals Δx and Δv and also recognizably the Product Rule.

Figure 3.1.4 Isaac Newton[8]

This argument is no better than Leibniz's as it relies heavily on the number $1/2$ to make it work. If we take any other increments in x and v whose total lengths are Δx and Δv it will simply not work. Try it and see.

In Newton's defense, he wasn't really trying to justify his mathematical methods in the Principia. His attention there was on physics, not math, so he was really just trying to give a convincing demonstration of his methods. You may decide for yourself how convincing his demonstration is.

Notice that there is no mention of limits of difference quotients or derivatives. In fact, the term derivative was not coined until 1797, by Lagrange. In a sense, these topics were not necessary at the time, as Leibniz and Newton both assumed that the curves they dealt with had tangent lines and, in fact, Leibniz explicitly used the tangent line to relate two differential quantities. This was consistent with the thinking of the time and for the duration of this chapter we will also assume that all quantities are differentiable. As we will see later this assumption leads to difficulties.

Both Newton and Leibniz were satisfied that their calculus provided answers that agreed with what was known at the time. For example $\mathrm{d}\left(x^2\right) = \mathrm{d}\left(xx\right) = x\,\mathrm{d}x + x\,\mathrm{d}x = 2x\,\mathrm{d}x$ and $\mathrm{d}\left(x^3\right) = \mathrm{d}\left(x^2x\right) = x^2\,\mathrm{d}x + x\,\mathrm{d}\left(x^2\right) = x^2 + x\left(2x\,\mathrm{d}x\right) = 3x^2\,\mathrm{d}x$, results that were essentially derived by others in different ways.

Problem 3.1.5

(a) Use Leibniz's product rule $\mathrm{d}\left(xv\right) = x\,\mathrm{d}v + v\,\mathrm{d}x$ to show that if n is a positive integer then $\mathrm{d}\left(x^n\right) = nx^{n-1}\,\mathrm{d}x$

(b) Use Leibniz's product rule to derive the quotient rule

$$\mathrm{d}\left(\frac{v}{y}\right) = \frac{y\ \mathrm{d}v - v\ \mathrm{d}y}{yy}.$$

[8] mathshistory.st-andrews.ac.uk/Biographies/Newton/

(c) Use the quotient rule to show that if $n$is a positive integer, then

$$\mathrm{d}\left(x^{-n}\right) = -nx^{-n-1}\,\mathrm{d}x.$$

□

Problem 3.1.6 Let p and q be integers with $q \neq 0$. Show that $\mathrm{d}\left(x^{\frac{p}{q}}\right) = \frac{p}{q}x^{\frac{p}{q}-1}\,\mathrm{d}x$ □

Leibniz also provided applications of his calculus to prove its worth. As an example he derived Snell's Law of Refraction from his calculus rules as follows.

Given that light travels through air at a speed of v_a and travels through water at a speed of v_w the problem is to find the fastest path from point A to point B.

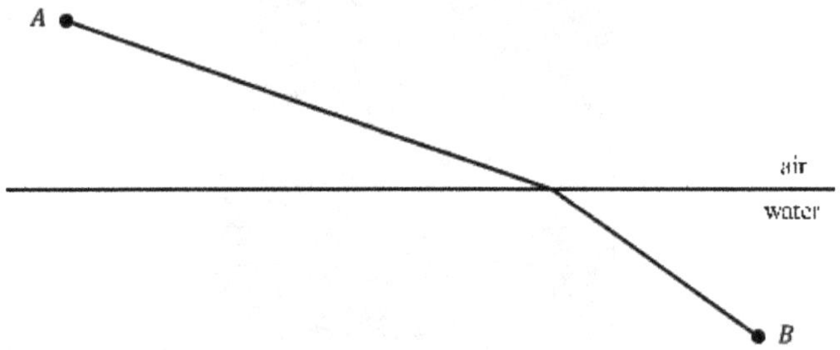

Figure 3.1.7

According to Fermat's Principle of Least Time, this fastest path is the one that light will travel.

Using the fact that Time = Distance / Velocity and the labeling in the picture below we can obtain a formula for the time T it takes for light to travel from A to B.

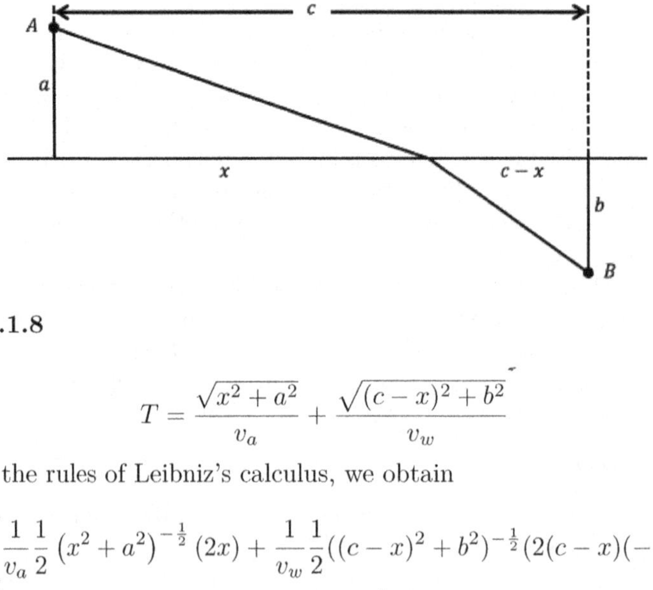

Figure 3.1.8

$$T = \frac{\sqrt{x^2+a^2}}{v_a} + \frac{\sqrt{(c-x)^2+b^2}}{v_w}$$

Using the rules of Leibniz's calculus, we obtain

$$\begin{aligned}\mathrm{d}T &= \left(\frac{1}{v_a}\frac{1}{2}\left(x^2+a^2\right)^{-\frac{1}{2}}(2x) + \frac{1}{v_w}\frac{1}{2}((c-x)^2+b^2)^{-\frac{1}{2}}(2(c-x)(-1))\right)\mathrm{d}x \\ &= \left(\frac{1}{v_a}\frac{x}{\sqrt{x^2+a^2}} - \frac{1}{v_w}\frac{c-x}{\sqrt{(c-x)^2+b^2}}\right)\mathrm{d}x.\end{aligned}$$

Using the fact that at the minimum value for T, $\mathrm{d}T = 0$, we have that the fastest path from $A$to B must satisfy $\frac{1}{v_a}\frac{x}{\sqrt{x^2+a^2}} = \frac{1}{v_w}\frac{c-x}{\sqrt{(c-x)^2+b^2}}$. Inserting the following angles

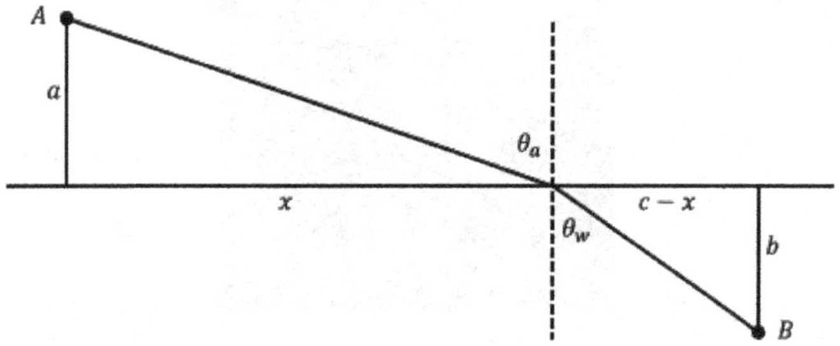

Figure 3.1.9

we get that the path that light travels must satisfy $\frac{\sin\theta_a}{v_a} = \frac{\sin\theta_w}{v_w}$ which is Snell's Law.

To compare 18th century and modern techniques we will consider Johann Bernoulli's solution of the Brachistochrone problem. In 1696, Bernoulli posed, and solved, the Brachistochrone problem; that is, to find the shape of a frictionless wire joining points A and B so that the time it takes for a bead to slide down under the force of gravity is as small as possible.

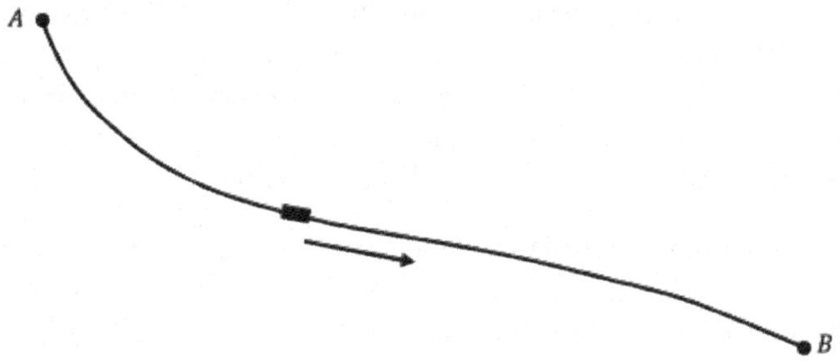

Figure 3.1.10

Bernoulli posed this "path of fastest descent" problem to challenge the mathematicians of Europe and used his solution to demonstrate the power of Leibniz's calculus as well as his own ingenuity.

> I, Johann Bernoulli, address the most brilliant mathematicians in the world. Nothing is more attractive to intelligent people than an honest, challenging problem, whose possible solution will bestow fame and remain as a lasting monument. Following the example set by Pascal, Fermat, etc., I hope to gain the gratitude of the whole scientific community by placing before the finest mathematicians of our time a problem which will test their methods and the strength of their intellect. If someone communicates to me the solution of the proposed problem, I shall publicly declare him worthy of praise. [11]

Figure 3.1.11 Johann Bernoulli[9]

In addition to Johann's, solutions were obtained from Newton, Leibniz, Johann's brother Jacob Bernoulli, and the Marquis de l'Hopital [15]. At the time there was an ongoing and very vitriolic controversy raging over whether Newton or Leibniz had been the first to invent calculus. An advocate of the methods of Leibniz, Bernoulli did not believe Newton would be able to solve the problem using his methods. Bernoulli attempted to embarrass Newton by sending him the problem. However Newton did solve it.

At this point in his life Newton had all but quit science and mathematics and was fully focused on his administrative duties as Master of the Mint. In part due to rampant counterfeiting, England's money had become severely devalued and the nation was on the verge of economic collapse. The solution was to recall all of the existing coins, melt them down, and strike new ones. As Master of the Mint this job fell to Newton [8]. As you might imagine this was a rather Herculean task. Nevertheless, according to his niece:

> When the problem in 1696 was sent by Bernoulli–Sir I.N. was in the midst of the hurry of the great recoinage and did not come home till four from the Tower very much tired, but did not sleep till he had solved it, which was by four in the morning.

He is later reported to have complained, "I do not love . . . to be . . . teezed by forreigners about Mathematical things" [2].

Newton submitted his solution anonymously, presumably to avoid more controversy. Nevertheless the methods used were so distinctively Newton's that Bernoulli is said to have exclaimed "*Tanquam ex ungue leonem.*"

Tanquam ex ungue leonem. "I know the lion by his claw."

Bernoulli's ingenious solution starts, interestingly enough, with Snell's Law of Refraction. He begins by considering the stratified medium in the following figure, where an object travels with velocities v_1, v_2, v_3, ... in the various layers.

[9] mathshistory.st-andrews.ac.uk/Biographies/Bernoulli_Johann/

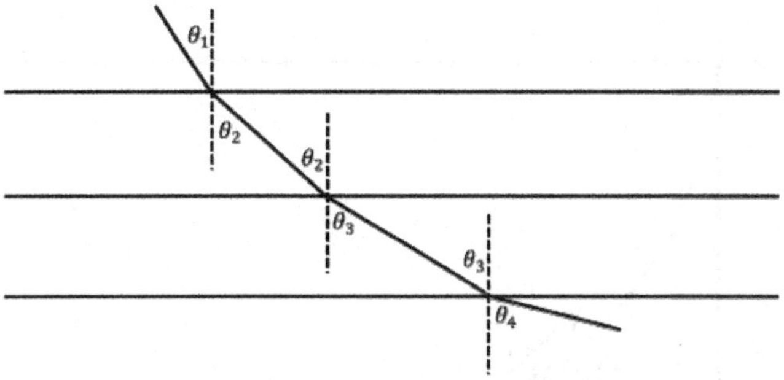

Figure 3.1.12

By repeatedly applying Snell's Law he concluded that the fastest path must satisfy

$$\frac{\sin\theta_1}{v_1} = \frac{\sin\theta_2}{v_2} = \frac{\sin\theta_3}{v_3} = \cdots.$$

In other words, the ratio of the sine of the angle that the curve makes with the vertical and the speed remains constant along this fastest path.

If we think of a continuously changing medium as stratified into infinitesimal layers and extend Snell's law to an object whose speed is constantly changing,

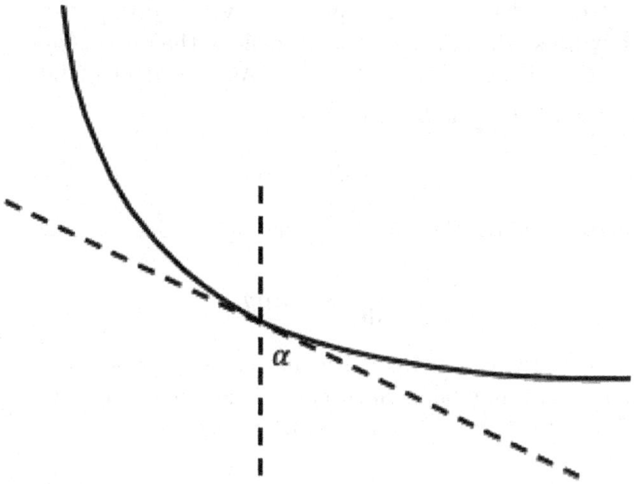

Figure 3.1.13

then along the fastest path, the ratio of the sine of the angle that the curve's tangent makes with the vertical, α, and the speed, v, must remain constant.

$$\frac{\sin\,\alpha}{v} = c.$$

If we include axes and let P denote the position of the bead at a particular time then we have the following picture.

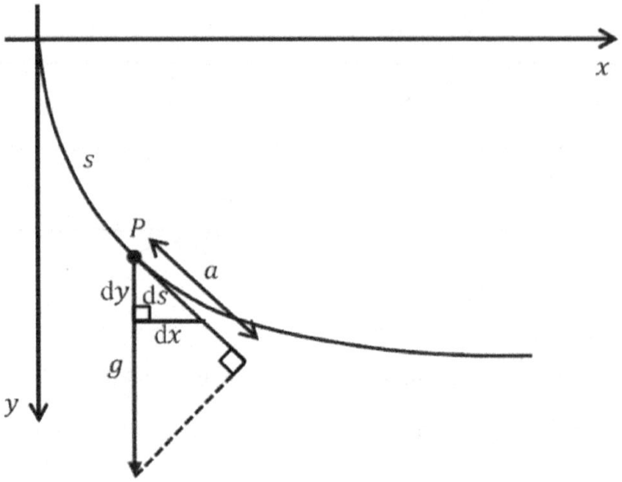

Figure 3.1.14

In the above figure, s denotes the length that the bead has traveled down to point P(that is, the arc length of the curve from the origin to that point) and a denotes the tangential component of the acceleration due to gravity g. Since the bead travels only under the influence of gravity then $\frac{\mathrm{d}v}{\mathrm{d}t} = a$.

To get a sense of how physical problems were approached using Leibniz's calculus we will use the above equation to show that $v = \sqrt{2gy}$.

By similar triangles we have $\frac{a}{g} = \frac{\mathrm{d}y}{\mathrm{d}s}$. As a student of Leibniz, Bernoulli would have regarded $\frac{\mathrm{d}y}{\mathrm{d}s}$ as a fraction so

$$a\,\mathrm{d}s = g\,\mathrm{d}y$$

and since acceleration is the rate of change of velocity we have

$$\frac{\mathrm{d}v}{\mathrm{d}t}\,\mathrm{d}s = g\,\mathrm{d}y.$$

Again, 18th century European mathematicians regarded $\mathrm{d}v$, $\mathrm{d}t$, and $\mathrm{d}s$ as infinitesimally small numbers which nevertheless obey all of the usual rules of algebra. Thus we can rearrange the above to get

$$\frac{\mathrm{d}s}{\mathrm{d}t}\,\mathrm{d}v = g\,\mathrm{d}y.$$

Since $\frac{\mathrm{d}s}{\mathrm{d}t}$ is the rate of change of position with respect to time it is, in fact, the velocity of the bead. That is

$$v\,\mathrm{d}v = g\,\mathrm{d}y.$$

Bernoulli would have interpreted this as a statement that two rectangles of height v and g, with respective widths $\mathrm{d}v$ and $\mathrm{d}y$ have equal area. Summing (integrating) all such rectangles we get:

$$\int v\,\mathrm{d}v = \int g\,\mathrm{d}y$$

$$\frac{v^2}{2} = gy$$

or

$$v = \sqrt{2gy}. \tag{I.4}$$

You are undoubtedly uncomfortable with the cavalier manipulation of infinitesimal quantities you've just witnessed, so we'll pause for a moment now to compare a modern development of equation (I.4) to Bernoulli's. As before we begin with the equation:

$$\frac{a}{g} = \frac{\mathrm{d}y}{\mathrm{d}s}$$
$$a = g\frac{\mathrm{d}y}{\mathrm{d}s}.$$

Moreover, since acceleration is the derivative of velocity this is the same as:

$$\frac{\mathrm{d}v}{\mathrm{d}t} = g\frac{\mathrm{d}y}{\mathrm{d}s}.$$

Now observe that by the Chain Rule $\frac{\mathrm{d}v}{\mathrm{d}t} = \frac{\mathrm{d}v}{\mathrm{d}s}\frac{\mathrm{d}s}{\mathrm{d}t}$. The physical interpretation of this formula is that velocity will depend on s, how far down the wire the bead has moved, but that the distance traveled will depend on how much time has elapsed. Therefore

$$\frac{\mathrm{d}v}{\mathrm{d}s}\frac{\mathrm{d}s}{\mathrm{d}t} = g\frac{\mathrm{d}y}{\mathrm{d}s}$$

or

$$\frac{\mathrm{d}s}{\mathrm{d}t}\frac{\mathrm{d}v}{\mathrm{d}s} = g\frac{\mathrm{d}y}{\mathrm{d}s}$$

and since $\frac{\mathrm{d}s}{\mathrm{d}t} = v$ we have

$$v\frac{\mathrm{d}v}{\mathrm{d}s} = g\frac{\mathrm{d}y}{\mathrm{d}s}$$

Integrating both sides with respect to s gives:

$$\int v\frac{\mathrm{d}v}{\mathrm{d}s}ds = g\int \frac{\mathrm{d}y}{\mathrm{d}s}ds$$
$$\int v dv = g\int dy$$

and integrating gives

$$\frac{v^2}{2} = gy$$

as before.

In effect, in the modern formulation we have traded the simplicity and elegance of differentials for a comparatively cumbersome repeated use of the Chain Rule. No doubt you noticed when taking Calculus that in the differential notation of Leibniz, the Chain Rule looks like "canceling" an expression in the top and bottom of a fraction: $\frac{\mathrm{d}y}{\mathrm{d}u}\frac{\mathrm{d}u}{\mathrm{d}x} = \frac{\mathrm{d}y}{\mathrm{d}x}$. This is because for 18th century mathematicians, this is exactly what it was.

To put it another way, 18th century mathematicians wouldn't have recognized a need for what we call the Chain Rule because this operation was a triviality for them. Just reduce the fraction. This begs the question: Why did we abandon such a clear, simple interpretation of our symbols in favor of the, comparatively, more cumbersome modern interpretation? This is one of the questions we will try to answer in this course.

Returning to the Brachistochrone problem we observe that

$$\frac{\sin\alpha}{v} = c$$

and since $\sin\alpha = \frac{dx}{ds}$ we see that

$$\begin{aligned} \frac{\frac{dx}{ds}}{\sqrt{2gy}} &= c \\ \frac{dx}{\sqrt{2gy(ds)^2}} &= c \\ \frac{dx}{\sqrt{2gy\left[(dx)^2 + (dy)^2\right]}} &= c. \end{aligned} \tag{I.5}$$

Bernoulli was then able to solve this differential equation.

Problem 3.1.15 Show that the equations $x = \frac{\phi - \sin\phi}{4gc^2}$, $y = \frac{1-\cos\phi}{4gc^2}$ satisfy equation (I.5). Bernoulli recognized this solution to be an inverted cycloid, the curve traced by a fixed point on a circle as the circle rolls along a horizontal surface. □

This illustrates the state of calculus in the late 1600's and early 1700's; the foundations of the subject were a bit shaky but there was no denying its power.

3.2 Power Series as Infinite Polynomials

Applied to polynomials, the rules of differential and integral calculus are straightforward. Indeed, differentiating and integrating polynomials represent some of the easiest tasks in a calculus course. For example, computing $\int (7 - x + x^2)\,\mathrm{d}x$ is relatively easy compared to computing $\int \sqrt[3]{1+x^3}\,\mathrm{d}x$. Unfortunately, not all functions can be expressed as a polynomial. For example, $f(x) = \sin x$ cannot be since a polynomial has only finitely many roots and the sine function has infinitely many roots, namely $\{n\pi \,|\, n \in \mathbb{Z}\}$. A standard technique in the 18th century was to write such functions as an "infinite polynomial," what we typically refer to as a power series. Unfortunately an "infinite polynomial" is a much more subtle object than a mere polynomial, which by definition is finite. For now we will not concern ourselves with these subtleties. We will follow the example of our forebears and manipulate all "polynomial-like" objects (finite or infinite) as if they are polynomials.

Definition 3.2.1 A **power series centered at a** is a series of the form

$$\sum_{n=0}^{\infty} a_n(x-a)^n = a_0 + a_1(x-a) + a_2(x-a)^2 + \cdots.$$

Often we will focus on the behavior of power series $\sum_{n=0}^{\infty} a_n x^n$, centered around 0, as the series centered around other values of a are obtained by shifting a series centered at 0. ♢

Before we continue, we will make the following notational comment. The most advantageous way to represent a series is using summation notation since there can be no doubt about the pattern to the terms. After all, this notation contains a formula for the general term. This being said, there are instances where writing this formula is not practical. In these cases, it is acceptable to write the sum by supplying the first few terms and using ellipses (the three dots). If this is done, then enough terms must be included to make the pattern clear to the reader.

Returning to our definition of a power series, consider, for example, the geometric series $\sum_{n=0}^{\infty} x^n = 1 + x + x^2 + \cdots$. If we multiply this series by $(1 - x)$, we obtain

$$(1-x)(1+x+x^2+\cdots) = (1+x+x^2+\cdots) - (x+x^2+x^3+\cdots) = 1.$$

This leads us to the power series representation

$$\frac{1}{1-x} = 1 + x + x^2 + \cdots = \sum_{n=0}^{\infty} x^n.$$

If we substitute $x = \frac{1}{10}$ into the above, we obtain

$$1 + \frac{1}{10} + \left(\frac{1}{10}\right)^2 + \left(\frac{1}{10}\right)^3 + \cdots = \frac{1}{1-\frac{1}{10}} = \frac{10}{9}.$$

This agrees with the fact that $.333\ldots = \frac{1}{3}$, and so $.111\ldots = \frac{1}{9}$, and $1.111\ldots = \frac{10}{9}$.

There are limitations to these formal manipulations however. Substituting $x = 1$ or $x = 2$ yields the questionable results

$$\frac{1}{0} = 1 + 1 + 1 + \cdots \quad \text{and} \quad \frac{1}{-1} = 1 + 2 + 2^2 + \cdots.$$

We are missing something important here, though it may not be clear exactly what. A series representation of a function works *sometimes*, but there are some problems. For now, we will continue to follow the example of our 18th century predecessors and ignore them. That is, for the rest of this section we will focus on the formal manipulations to obtain and use power series representations of various functions. Keep in mind that this is all highly suspect until we can resolve problems like those just given.

Power series became an important tool in analysis in the 1700's. By representing various functions as power series they could be dealt with as if they were (infinite) polynomials. The following is an example.

Example 3.2.2 Solve the following Initial Value problem: Find $y(x)$ given that $\frac{dy}{dx} = y$, $y(0) = 1$.

Comment. A few seconds of thought should convince you that the solution of this problem is $y(x) = e^x$. We will ignore this for now in favor of emphasising the technique.

Assuming the solution can be expressed as a power series we have

$$y = \sum_{n=0}^{\infty} a_n x^n = a_0 + a_1 x + a_2 x^2 + \cdots.$$

Differentiating gives us

$$\frac{dy}{dx} = a_1 + 2a_2 x + 3a_3 x^2 + 4a_4 x^3 + \ldots.$$

Since $\frac{dy}{dx} = y$ we see that

$$a_1 = a_0\,,\ 2a_2 = a_1\,,\ 3a_3 = a_2\,,\ \ldots,\ na_n = a_{n-1}\,,\ldots.$$

This leads to the relationship

$$a_n = \frac{1}{n}a_{n-1} = \frac{1}{n(n-1)}a_{n-2} = \cdots = \frac{1}{n!}a_0.$$

Thus the series solution of the differential equation is

$$y = \sum_{n=0}^{\infty} \frac{a_0}{n!}x^n = a_0 \sum_{n=0}^{\infty} \frac{1}{n!}x^n.$$

Using the initial condition $y(0) = 1$, we get $1 = a_0(1 + 0 + \frac{1}{2!}0^2 + \cdots) = a_0$. Thus the solution to the initial problem is $y = \sum_{n=0}^{\infty} \frac{1}{n!}x^n$. Let's call this function $E(x)$. Then by definition

$$E(x) = \sum_{n=0}^{\infty} \frac{1}{n!}x^n = 1 + \frac{x^1}{1!} + \frac{x^2}{2!} + \frac{x^3}{3!} + \cdots.$$

□

Let's examine some properties of this function. The first property is clear from the definition.

Property 1. $E(0) = 1$

Property 2. $E(x + y) = E(x)E(y)$.

To see this we multiply the two series together, so we have

$$\begin{aligned}
E(x)E(y) &= \left(\sum_{n=0}^{\infty} \frac{1}{n!}x^n\right)\left(\sum_{n=0}^{\infty} \frac{1}{n!}y^n\right) \\
&= \left(\frac{x^0}{0!} + \frac{x^1}{1!} + \frac{x^2}{2!} + \frac{x^3}{3!} + \ldots\right)\left(\frac{y^0}{0!} + \frac{y^1}{1!} + \frac{y^2}{2!} + \frac{y^3}{3!} + \ldots\right) \\
&= \frac{x^0}{0!}\frac{y^0}{0!} + \frac{x^0}{0!}\frac{y^1}{1!} + \frac{x^1}{1!}\frac{y^0}{0!} + \frac{x^0}{0!}\frac{y^2}{2!} + \frac{x^1}{1!}\frac{y^1}{1!} + \frac{x^2}{2!}\frac{y^0}{0!} \\
&\quad + \frac{x^0}{0!}\frac{y^3}{3!} + \frac{x^1}{1!}\frac{y^2}{2!} + \frac{x^2}{2!}\frac{y^1}{1!} + \frac{x^3}{3!}\frac{y^0}{0!} + \ldots \\
&= \frac{x^0}{0!}\frac{y^0}{0!} + \left(\frac{x^0}{0!}\frac{y^1}{1!} + \frac{x^1}{1!}\frac{y^0}{0!}\right) \\
&\quad + \left(\frac{x^0}{0!}\frac{y^2}{2!} + \frac{x^1}{1!}\frac{y^1}{1!} + \frac{x^2}{2!}\frac{y^0}{0!}\right) \\
&\quad + \left(\frac{x^0}{0!}\frac{y^3}{3!} + \frac{x^1}{1!}\frac{y^2}{2!} + \frac{x^2}{2!}\frac{y^1}{1!} + \frac{x^3}{3!}\frac{y^0}{0!}\right) + \ldots \\
&= \frac{1}{0!} + \frac{1}{1!}\left(\frac{1!}{0!1!}x^0y^1 + \frac{1!}{1!0!}x^1y^0\right) \\
&\quad + \frac{1}{2!}\left(\frac{2!}{0!2!}x^0y^2 + \frac{2!}{1!1!}x^1y^1 + \frac{2!}{2!0!}x^2y^0\right) \\
&\quad + \frac{1}{3!}\left(\frac{3!}{0!3!}x^0y^3 + \frac{3!}{1!2!}x^1y^2 + \frac{3!}{2!1!}x^2y^1 + \frac{3!}{3!0!}x^3y^0\right) + \ldots
\end{aligned}$$

$$\begin{aligned}
E(x)E(y) &= \frac{1}{0!} + \frac{1}{1!}\left(\binom{1}{0}x^0y^1 + \binom{1}{1}x^1y^0\right) \\
&\quad + \frac{1}{2!}\left(\binom{2}{0}x^0y^2 + \binom{2}{1}x^1y^1 + \binom{2}{2}x^2y^0\right) \\
&\quad + \frac{1}{3!}\left(\binom{3}{0}x^0y^3 + \binom{3}{1}x^1y^2 + \binom{3}{2}x^2y^1 + \binom{3}{3}x^3y^0\right) + \ldots
\end{aligned}$$

$$= \frac{1}{0!} + \frac{1}{1!}(x+y)^1 + \frac{1}{2!}(x+y)^2 + \frac{1}{3!}(x+y)^3 + \dots$$
$$= E(x+y). \tag{I.6}$$

Property 3. If m is a positive integer then $E(mx) = (E(x))^m$. In particular, $E(m) = (E(1))^m$.

Problem 3.2.3 Prove Property 3. □

Property 4. $E(-x) = \frac{1}{E(x)} = (E(x))^{-1}$.

Problem 3.2.4 Prove Property 4. □

Property 5. If n is an integer with $n \neq 0$, then $E(\frac{1}{n}) = \sqrt[n]{E(1)} = (E(1))^{1/n}$.

Problem 3.2.5 Prove Property 5. □

Property 6. If m and n are integers with $n \neq 0$, then $E\left(\frac{m}{n}\right) = (E(1))^{m/n}$.

Problem 3.2.6 Prove Property 6. □

Definition 3.2.7 Let $E(1)$ be denoted by the number e. Using the series $e = E(1) = \sum_{n=0}^{\infty} \frac{1}{n!}$, we can approximate e to any degree of accuracy. In particular $e \approx 2.71828$. ♢

In light of Property 6, we see that for any rational number r, $E(r) = e^r$. Not only does this give us the series representation $e^r = \sum_{n=0}^{\infty} \frac{1}{n!} r^n$ for any rational number r, but it gives us a way to define e^x for irrational values of x as well. That is, we can define

$$e^x = E(x) = \sum_{n=0}^{\infty} \frac{1}{n!} x^n$$

for any real number x.

As an illustration, we now have $e^{\sqrt{2}} = \sum_{n=0}^{\infty} \frac{1}{n!} \left(\sqrt{2}\right)^n$. The expression $e^{\sqrt{2}}$ is meaningless if we try to interpret it as one irrational number raised to another. What does it mean to raise anything to the $\sqrt{2}$ power? However the series $\sum_{n=0}^{\infty} \frac{1}{n!} \left(\sqrt{2}\right)^n$ does seem to have meaning and it can be used to extend the exponential function to irrational exponents. In fact, defining the exponential function via this series answers the question we raised in Chapter 2: What does $4^{\sqrt{2}}$ mean?

It means $4^{\sqrt{2}} = e^{\sqrt{2}\log 4} = \sum_{n=0}^{\infty} \frac{(\sqrt{2}\log 4)^n}{n!}$.

This may seem to be the long way around just to define something as simple as exponentiation. But this is a fundamentally misguided attitude. Exponentiation only *seems* simple because we've always thought of it as repeated multiplication (in $\mathbb{Z}$) or root-taking (in $\mathbb{Q}$). When we expand the operation to the real numbers this simply can't be the way we interpret something like $4^{\sqrt{2}}$. How do you take the product of $\sqrt{2}$ copies of 4? The concept is meaningless. What we need is an interpretation of $4^{\sqrt{2}}$ which is consistent with, say $4^{3/2} = \left(\sqrt{4}\right)^3 = 8$. This is exactly what the series representation of e^x provides.

We also have a means of computing integrals as series. For example, the famous "bell shaped" curve given by the function $f(x) = \frac{1}{\sqrt{2\pi}} e^{-\frac{x^2}{2}}$ is of vital importance in statistics and must be integrated to calculate probabilities. The power series we developed gives us a method of integrating this function. For example, we have

$$\begin{aligned}\int_{x=0}^{b}\frac{1}{\sqrt{2\pi}}e^{-\frac{x^2}{2}}dx &= \frac{1}{\sqrt{2\pi}}\int_{x=0}^{b}\left(\sum_{n=0}^{\infty}\frac{1}{n!}\left(\frac{-x^2}{2}\right)^n\right)dx\\ &= \frac{1}{\sqrt{2\pi}}\sum_{n=0}^{\infty}\left(\frac{(-1)^n}{n!2^n}\int_{x=0}^{b}x^{2n}dx\right)\\ &= \frac{1}{\sqrt{2\pi}}\sum_{n=0}^{\infty}\left(\frac{(-1)^n b^{2n+1}}{n!2^n(2n+1)}\right).\end{aligned}$$

This series can be used to approximate the integral to any degree of accuracy. The ability to provide such calculations made power series of paramount importance in the 1700's.

Problem 3.2.8

(a) Show that if $y=\sum_{n=0}^{\infty}a_nx^n$ satisfies the differential equation $\frac{d^2y}{dx^2}=-y$, then

$$a_{n+2}=\frac{-1}{(n+2)(n+1)}a_n$$

and conclude that

$$y=a_0+a_1x-\frac{1}{2!}a_0x^2-\frac{1}{3!}a_1x^3+\frac{1}{4!}a_0x^4+\frac{1}{5!}a_1x^5-\frac{1}{6!}a_0x^6-\frac{1}{7!}a_1x^7+\cdots.$$

(b) Since $y=\sin x$ satisfies $\frac{d^2y}{dx^2}=-y$, we see that

$$\sin x=a_0+a_1x-\frac{1}{2!}a_0x^2-\frac{1}{3!}a_1x^3+\frac{1}{4!}a_0x^4+\frac{1}{5!}a_1x^5-\frac{1}{6!}a_0x^6-\frac{1}{7!}a_1x^7+\cdots$$

for some constants a_0 and a_1. Show that in this case $a_0=0$ and $a_1=1$ and obtain

$$\sin x=x-\frac{1}{3!}x^3+\frac{1}{5!}x^5-\frac{1}{7!}x^7+\cdots=\sum_{n=0}^{\infty}\frac{(-1)^n}{(2n+1)!}x^{2n+1}.$$

□

Problem 3.2.9

(a) Use the series

$$\sin x=x-\frac{1}{3!}x^3+\frac{1}{5!}x^5-\frac{1}{7!}x^7+\cdots=\sum_{n=0}^{\infty}\frac{(-1)^n}{(2n+1)!}x^{2n+1}$$

to obtain the series

$$\cos x=1-\frac{1}{2!}x^2+\frac{1}{4!}x^4-\frac{1}{6!}x^6+\cdots=\sum_{n=0}^{\infty}\frac{(-1)^n}{(2n)!}x^{2n}.$$

(b) Let $s(x,N)=\sum_{n=0}^{N}\frac{(-1)^n}{(2n+1)!}x^{2n+1}$ and $c(x,N)=\sum_{n=0}^{N}\frac{(-1)^n}{(2n)!}x^{2n}$ and use a computer algebra system to plot these for $-4\pi\le x\le 4\pi$, $N=1,2,5,10,15$. Describe what is happening to the series as N becomes larger.

□

Problem 3.2.10 Use the geometric series, $\frac{1}{1-x} = 1 + x + x^2 + x^3 + \cdots = \sum_{n=0}^{\infty} x^n$, to obtain a series for $\frac{1}{1+x^2}$ and use this to obtain the series

$$\arctan x = x - \frac{1}{3}x^3 + \frac{1}{5}x^5 - \cdots = \sum_{n=0}^{\infty}(-1)^n \frac{1}{2n+1}x^{2n+1}.$$

Use the series above to obtain the series $\frac{\pi}{4} = \sum_{n=0}^{\infty}(-1)^n \frac{1}{2n+1}$. □

The series for arctangent was known by James Gregory (1638-1675) and it is sometimes referred to as "Gregory's series." Leibniz independently discovered $\frac{\pi}{4} = 1 - \frac{1}{3} + \frac{1}{5} - \frac{1}{7} + \cdots$ by examining the area of a circle. Though it gives us a means for approximating π to any desired accuracy, the series converges too slowly to be of any practical use. For example, if we compute the sum of the first 1000 terms we get

$$4\left(\sum_{n=0}^{1000}(-1)^n \frac{1}{2n+1}\right) \approx 3.142591654$$

which only approximates π to two decimal places.

Newton knew of these results and the general scheme of using series to compute areas under curves. These results motivated Newton to provide a series approximation for π as well, which, hopefully, would converge faster. We will use modern terminology to streamline Newton's ideas. First notice that $\frac{\pi}{4} = \int_{x=0}^{1} \sqrt{1-x^2}\, dx$ as this integral gives the area of one quarter of the unit circle. The trick now is to find series that represents $\sqrt{1-x^2}$.

To this end we start with the binomial theorem

$$(a+b)^N = \sum_{n=0}^{N} \binom{N}{n} a^{N-n} b^n,$$

where

$$\begin{aligned}\binom{N}{n} &= \frac{N!}{n!\,(N-n)!} \\ &= \frac{N\,(N-1)\,(N-2)\cdots(N-n+1)}{n!} \\ &= \frac{\prod_{j=0}^{n-1}(N-j)}{n!}.\end{aligned}$$

Unfortunately, we now have a small problem with our notation which will be a source of confusion later if we don't fix it. So we will pause to address this matter. We will come back to the binomial expansion afterward.

This last expression is becoming awkward in much the same way that an expression like

$$1 + \frac{1}{2} + \left(\frac{1}{2}\right)^2 + \left(\frac{1}{2}\right)^3 + \ldots + \left(\frac{1}{2}\right)^k$$

is awkward. Just as this sum is less cumbersome when written as $\sum_{n=0}^{k}\left(\frac{1}{2}\right)^n$ the *product*

$$N\,(N-1)\,(N-2)\cdots(N-n+1)$$

is less cumbersom when we write it as $\prod_{j=0}^{n-1}(N-j)$.

A capital pi (Π) is used to denote a product in the same way that a capital sigma (Σ) is used to denote a sum. The most familiar example would be writing

$$n! = \prod_{j=1}^{n} j.$$

Just as it is convenient to define $0! = 1$, we will find it convenient to define $\prod_{j=1}^{0} = 1$. Similarly, the fact that $\binom{N}{0} = 1$ leads to the convention $\prod_{j=0}^{-1}(N-j) = 1$. Strange as this may look, it is convenient and *is* consistent with the convention $\sum_{j=0}^{-1} s_j = 0$.

Returning to the binomial expansion and recalling our convention

$$\prod_{j=0}^{-1}(N-j) = 1,$$

we can write,

$$(1+x)^N = 1 + \sum_{n=1}^{N}\left(\frac{\prod_{j=0}^{n-1}(N-j)}{n!}\right)x^n = \sum_{n=0}^{N}\left(\frac{\prod_{j=0}^{n-1}(N-j)}{n!}\right)x^n.$$

These two representations probably look the same at first. Take a moment and be sure you see where they differ.

There is an advantage to using this convention (especially when programing a product into a computer), but this is not a deep mathematical insight. It is just a notational convenience and we don't want you to fret over it, so we will use both formulations (at least initially).

Notice that we can extend the above definition of $\binom{N}{n}$ to values $n > N$. In this case, $\prod_{j=0}^{n-1}(N-j)$ will equal 0 as one of the factors in the product will be 0 (the one where $n = N$). This gives us that $\binom{N}{n} = 0$ when $n > N$ and so

$$(1+x)^N = 1 + \sum_{n=1}^{\infty}\left(\frac{\prod_{j=0}^{n-1}(N-j)}{n!}\right)x^n = \sum_{n=0}^{\infty}\left(\frac{\prod_{j=0}^{n-1}(N-j)}{n!}\right)x^n$$

holds true for any nonnegative integer N. Essentially Newton asked if it could be possible that the above equation could hold values of N which are not nonnegative integers. For example, if the equation held true for $N = \frac{1}{2}$, we would obtain

$$(1+x)^{\frac{1}{2}} = 1 + \sum_{n=1}^{\infty}\left(\frac{\prod_{j=0}^{n-1}\left(\frac{1}{2}-j\right)}{n!}\right)x^n = \sum_{n=0}^{\infty}\left(\frac{\prod_{j=0}^{n-1}\left(\frac{1}{2}-j\right)}{n!}\right)x^n$$

or

$$(1+x)^{\frac{1}{2}} = 1 + \frac{1}{2}x + \frac{\frac{1}{2}\left(\frac{1}{2}-1\right)}{2!}x^2 + \frac{\frac{1}{2}\left(\frac{1}{2}-1\right)\left(\frac{1}{2}-2\right)}{3!}x^3 + \cdots. \tag{I.7}$$

Notice that since $1/2$ is not an integer the series no longer terminates. Although Newton did not prove that this series was correct (nor did we), he tested it by multiplying the series by itself. When he saw that by squaring the series he started to obtain $1 + x + 0\,x^2 + 0\,x^3 + \cdots$, he was convinced that the series was exactly equal to $\sqrt{1+x}$.

Problem 3.2.11

Consider the series representation

$$\begin{aligned}(1+x)^{\frac{1}{2}} &= 1 + \sum_{n=1}^{\infty}\frac{\prod_{j=0}^{n-1}\left(\frac{1}{2}-j\right)}{n!}x^n \\ &= \sum_{n=0}^{\infty}\frac{\prod_{j=0}^{n-1}\left(\frac{1}{2}-j\right)}{n!}x^n.\end{aligned}$$

Multiply this series by itself and compute the coefficients for x^0, x^1, x^2, x^3, x^4 in the resulting series. □

Problem 3.2.12 Let

$$S(x,M)=\sum_{n=0}^{M}\frac{\prod_{j=0}^{n-1}\left(\frac{1}{2}-j\right)}{n!}x^n.$$

Use a computer algebra system to plot $S(x,M)$ for $M=5, 10, 15, 95, 100$ and compare these to the graph for $\sqrt{1+x}$. What seems to be happening? For what values of x does the series appear to converge to $\sqrt{1+x}$? □

Convinced that he had the correct series, Newton used it to find a series representation of $\int_{x=0}^{1}\sqrt{1-x^2}\,\mathrm{d}x$.

Problem 3.2.13 Use the series $(1+x)^{\frac{1}{2}}=\sum_{n=0}^{\infty}\frac{\prod_{j=0}^{n-1}\left(\frac{1}{2}-j\right)}{n!}x^n$ to obtain the series

$$\begin{aligned}\frac{\pi}{4}&=\int_{x=0}^{1}\sqrt{1-x^2}\,\mathrm{d}x\\&=\sum_{n=0}^{\infty}\left(\frac{\prod_{j=0}^{n-1}\left(\frac{1}{2}-j\right)}{n!}\right)\left(\frac{(-1)^n}{2n+1}\right)\\&=1-\frac{1}{6}-\frac{1}{40}-\frac{1}{112}-\frac{5}{1152}-\cdots.\end{aligned}$$

Use a computer algebra system to sum the first 100 terms of this series and compare the answer to $\frac{\pi}{4}$. □

Again, Newton had a series which could be verified (somewhat) computationally. This convinced him even further that he had the correct series.

Problem 3.2.14

a Show that

$$\int_{x=0}^{1/2}\sqrt{x-x^2}\,\mathrm{d}x=\sum_{n=0}^{\infty}\frac{(-1)^n\prod_{j=0}^{n-1}\left(\frac{1}{2}-j\right)}{\sqrt{2}\,n!\,(2n+3)\,2^n}$$

and use this to show that

$$\pi=16\left(\sum_{n=0}^{\infty}\frac{(-1)^n\prod_{j=0}^{n-1}\left(\frac{1}{2}-j\right)}{\sqrt{2}\,n!\,(2n+3)\,2^n}\right).$$

b We now have two series for calculating π : the one from part (a) and the one derived earlier, namely

$$\pi=4\left(\sum_{n=0}^{\infty}\frac{(-1)^n}{2n+1}\right).$$

We will explore which one converges to π faster. With this in mind, define $S1(N)=16\left(\sum_{n=0}^{N}\frac{(-1)^n\prod_{j=0}^{n-1}\left(\frac{1}{2}-j\right)}{\sqrt{2}\,n!(2n+3)2^n}\right)$ and $S2(N)=4\left(\sum_{n=0}^{N}\frac{(-1)^n}{2n+1}\right)$. Use a computer algebra system to compute $S1(N)$and $S2(N)$ for $N=5, 10, 15, 20$. Which one appears to converge to π faster?

□

In general the series representation

$$(1+x)^{\alpha} = \sum_{n=0}^{\infty}\left(\frac{\prod_{j=0}^{n-1}(\alpha-j)}{n!}\right)x^n$$
$$= 1+\alpha x+\frac{\alpha(\alpha-1)}{2!}x^2+\frac{\alpha(\alpha-1)(\alpha-2)}{3!}x^3+\cdots$$

is called the **binomial series** (or Newton's binomial series). This series is correct when α is a non-negative integer (after all, that is how we got the series). We can also see that it is correct when $\alpha=-1$ as we obtain

$$(1+x)^{-1} = \sum_{n=0}^{\infty}\left(\frac{\prod_{j=0}^{n-1}(-1-j)}{n!}\right)x^n$$
$$= 1+(-1)x+\frac{-1(-1-1)}{2!}x^2+\frac{-1(-1-1)(-1-2)}{3!}x^3+\cdots$$
$$= 1-x+x^2-x^3+\cdots$$

which can be obtained from the geometric series $\frac{1}{1-x}=1+x+x^2+\cdots$.

In fact, the binomial series is the correct series representation for all values of the exponent α (though we haven't proved this yet).

Problem 3.2.15 Let k be a positive integer. Find the power series, centered at zero, for $f(x)=(1-x)^{-k}$ by

(a) Differentiating the geometric series $(k-1)$ times.

(b) Applying the binomial series.

(c) Compare these two results.

□

Figure 3.2.16 Leonhard Euler[10]

Leonhard Euler was a master at exploiting power series. In 1735, the 28 year-old Euler won acclaim for what is now called the Basel problem: to find a closed form for $\sum_{n=1}^{\infty}\frac{1}{n^2}$. Other mathematicans knew that the series converged, but Euler was the first to find its exact value. The following problem essentially provides Euler's solution.

Problem 3.2.17 The Basel Problem.

(a) Show that the power series for $\frac{\sin x}{x}$ is given by $1-\frac{1}{3!}x^2+\frac{1}{5!}x^4-\cdots$

[10]mathshistory.st-andrews.ac.uk/Biographies/Euler/

(b) Use (a) to infer that the roots of $1 - \frac{1}{3!}x^2 + \frac{1}{5!}x^4 - \cdots$ are given by

$$x = \pm\pi,\ \pm 2\pi,\ \pm 3\pi,\ \ldots$$

(c) Suppose $p(x) = a_0 + a_1 x + \cdots + a_n x^n$ is a polynomial with roots $r_1, r_2, \ldots, r_n$. Show that if $a_0 \neq 0$, then all the roots are non-zero and

$$p(x) = a_0 \left(1 - \frac{x}{r_1}\right)\left(1 - \frac{x}{r_2}\right)\cdots\left(1 - \frac{x}{r_n}\right).$$

(d) Assuming that the result in part (c) holds for an infinite polynomial (power series), deduce that

$$1 - \frac{1}{3!}x^2 + \frac{1}{5!}x^4 - \cdots = \left(1 - \left(\frac{x}{\pi}\right)^2\right)\left(1 - \left(\frac{x}{2\pi}\right)^2\right)\left(1 - \left(\frac{x}{3\pi}\right)^2\right)\cdots$$

(e) Expand this product to deduce

$$\sum_{n=1}^{\infty} \frac{1}{n^2} = \frac{\pi^2}{6}.$$

□

Problem 3.2.18 Euler's Formula.

(a) Use the power series expansion of e^x, $\sin x$, and $\cos x$ to derive **Euler's Formula**:

$$e^{i\theta} = cos\theta + i\sin\theta.$$

(b) Use Euler's formula to derive the Addition/Subtraction formulas from Trigonometry:

$$\sin(\alpha \pm \beta) = \sin\alpha\cos\beta \pm \sin\beta\cos\alpha$$

$$\cos(\alpha \pm \beta) = \cos\alpha\cos\beta \mp \sin\alpha\sin\beta$$

(c) Use Euler's formula to show that

$$\sin 2\theta = 2\cos\theta\sin\theta$$

$$\cos 2\theta = \cos^2\theta - \sin^2\theta$$

(d) Use Euler's formula to show that

$$\sin 3\theta = 3\cos^2\theta\sin\theta - \sin^3\theta$$

$$\cos 3\theta = \cos^3\theta - \cos\theta\sin^2\theta$$

(e) Find a formula $\sin(n\theta)$ and $\cos(n\theta)$ for any positive integer n.

□

3.3 Additional Problems

Problem 3.3.1 Use the geometric series to obtain the series

$$\begin{aligned}\ln(1+x) &= x - \frac{1}{2}x^2 + \frac{1}{3}x^3 - \cdots \\ &= \sum_{n=0}^{\infty} \frac{(-1)^n}{n+1} x^{n+1}.\end{aligned}$$

□

Problem 3.3.2 *Without* using Taylor's Theorem, represent the following functions as power series expanded about 0 (i.e., in the form $\sum_{n=0}^{\infty} a_n x^n$).

(a) $\ln(1-x^2)$

(b) $\frac{x}{1+x^2}$

(c) $\arctan(x^3)$

(d) $\ln(2+x)$

Hint.

$2+x = 2\left(1+\frac{x}{2}\right)$

□

Problem 3.3.3 Let a be a positive real number. Find a power series for a^x expanded about 0.

Hint.

$a^x = e^{\ln(a^x)}$ □

Problem 3.3.4 Represent the function $\sin x$ as a power series expanded about a (i.e., in the form $\sum_{n=0}^{\infty} a_n (x-a)^n$).

Hint.

$\sin x = \sin(a + x - a)$. □

Problem 3.3.5 *Without* using Taylor's Theorem, represent the following functions as a power series expanded about a for the given value of a (i.e., in the form $\sum_{n=0}^{\infty} a_n (x-a)^n$).

(a) $\ln x$, $a = 1$

(b) e^x, $a = 3$

(c) $x^3 + 2x^2 + 3$, $a = 1$

(d) $\frac{1}{x}$, $a = 5$

□

Problem 3.3.6 Evaluate the following integrals as series.

(a) $\displaystyle\int_{x=0}^{1} e^{x^2}\,dx$

(b) $\displaystyle\int_{x=0}^{1} \frac{1}{1+x^4}\,dx$

(c) $\displaystyle\int_{x=0}^{1} \sqrt[3]{1-x^3}\,dx$

□

Chapter 4

Questions Concerning Power Series

4.1 Taylor's Formula

As we saw in the previous chapter, representing functions as power series was a fruitful strategy for mathematicans in the eighteenth century (as it still is). Differentiating and integrating power series term by term was relatively easy, *seemed* to work, and led to many applications. Furthermore, power series representations for all of the elementary functions could be obtained if one was clever enough.

However, cleverness is an unreliable tool. Is there some systematic way to find a power series for a given function? To be sure, there were nagging questions: If we can find a power series, how do we know that the series we've created represents the function we started with? Even worse, is it possible for a function to have more than one power series representation centered at a given value a? This uniqueness issue is addressed by the following theorem.

Theorem 4.1.1 *If* $f(x) = \sum_{n=0}^{\infty} a_n(x-a)^n$, *then* $a_n = \frac{f^{(n)}(a)}{n!}$, *where* $f^{(n)}(a)$ *represents the* n^{th} *derivative of* f *evaluated at* a.

A few comments about Theorem 4.1.1 are in order. Notice that we did *not* start with a function and derive its series representation. Instead we *defined* $f(x)$ to be the series we wrote down. This assumes that the expression $\sum_{n=0}^{\infty} a_n(x-a)^n$ actually has meaning (that it converges). At this point we have every reason to expect that it does, however expectation is not proof so we note that this is an assumption, not an established truth. Similarly, the idea that we can differentiate an infinite polynomial term-by-term as we would a finite polynomial is also assumed. As before, we follow in the footsteps of our 18th century forebears in making these assumptions. For now.

Problem 4.1.2 Prove Theorem 4.1.1.

Hint.

$f(a) = a_0 + a_1(a-a) + a_2(a-a)^2 + \cdots = a_0$. Differentiate to obtain the other terms. □

From Theorem 4.1.1 we see that if we do start with the function $f(x)$ then no matter how we obtain its power series, the result will always be the same. The series

$$\sum_{n=0}^{\infty} \frac{f^{(n)}(a)}{n!}(x-a)^n = f(a) + f'(a)(x-a) + \frac{f''(a)}{2!}(x-a)^2 + \frac{f'''(a)}{3!}(x-a)^3 + \cdots$$

Figure 4.1.3 Brook Taylor[11]

is called the ***Taylor series*** *for f expanded about (centered at) a.* Although this systematic "machine" for obtaining power series for a function seems to have been known to a number of mathematicians in the early 1700's, Brook Taylor was the first to publish this result in his *Methodus Incrementorum* (1715). The special case when $a = 0$ was included by Colin Maclaurin in his *Treatise of Fluxions* (1742). Thus when $a = 0$, the series $\sum_{n=0}^{\infty} \frac{f^{(n)}(0)}{n!} x^n$ is often called the *Maclaurin Series for f*.

The "prime notation" for the derivative was not used by Taylor, Maclaurin or their contemporaries. It was introduced by Joseph Louis Lagrange in his 1779 work *Théorie des Fonctions Analytiques.* In that work, Lagrange sought to get rid of Leibniz's infinitesimals and base calculus on the power series idea. His idea was that by representing every function as a power series, calculus could be done "algebraically" by manipulating power series and examining various aspects of the series representation instead of appealing to the "controversial" notion of infinitesimals. He implicitly assumed that every continuous function could be replaced with its power series representation.

That is, he wanted to think of the Taylor series as a "great big polynomial," because polynomials are easy to work with. It was a very simple, yet exceedingly clever and far-reaching idea. Since $e^x = 1 + x + x^2/2 + \ldots$, for example, why not just define the exponential to be the series and work with the series. After all, the series is just a very long polynomial.

This idea did not come out of nowhere. Leonhard Euler had put exactly that idea to work to solve many problems throughout the 18th century. Some of his solutions are still quite breath-taking when you first see them [14].

Taking his cue from the Taylor series

$$f(x) = \sum_{n=0}^{\infty} \frac{f^{(n)}(a)}{n!}(x - a)^n$$

[11] mathshistory.st-andrews.ac.uk/Biographies/Taylor/

Figure 4.1.4 Joseph-Louis Lagrange[12]

Lagrange observed that the coefficient of $(x-a)^n$ provides the derivative of f at a (divided by $n!$). Modifying the formula above to suit his purpose, Lagrange supposed that every differentiable function could be represented as

$$f(x) = \sum_{n=0}^{\infty} g_n(a)(x-a)^n.$$

If we regard the parameter a as a variable then g_1 is the derivative of f, $g_2 = 2f''$ and generally

$$g_n = n!f^{(n)}.$$

Lagrange dubbed his function g_1 the *"fonction dérivée"* from which we get the modern name "derivative."

All in all, this was a very clever and insightful idea whose only real flaw is that its fundamental assumption is not true. It turns out that not every differentiable function can be represented as a Taylor series. This was demonstrated very dramatically by Augustin Cauchy's famous counter-example

$$f(x) = \begin{cases} e^{-\frac{1}{x^2}} & x \neq 0 \\ 0 & x = 0 \end{cases}. \tag{I.1}$$

This function is actually infinitely differentiable everywhere but its Maclaurin series (that is, a Taylor series with $a = 0$) does not converge to f because all of its derivatives at the origin are equal to zero: $f^{(n)}(0) = 0, \forall n \in \mathbb{N}$.

Computing these derivatives using the definition you learned in calculus is not conceptually difficult but the formulas involved do become complicated rather quickly. Some care must be taken to avoid error.

To begin with, let's compute a few derivatives when $x \neq 0$.

$$\begin{aligned} f^{(0)}(x) &= e^{x^{-2}} \\ f^{(1)}(x) &= 2x^{-3}e^{-x^{-2}} \\ f^{(2)}(x) &= \left(4x^{-6} - 6x^{-4}\right)e^{-x^{-2}}. \end{aligned}$$

As you can see the calculations are already getting a little complicated and we've only taken the second derivative. To streamline things a bit we take $y = x^{-1}$, and define $p_2(x) = 4x^6 - 6x^4$ so that

$$f^{(2)}(x) = p_2(x^{-1})e^{-x^{-2}} = p_2(y)e^{-y^2}.$$

[12] mathshistory.st-andrews.ac.uk/Biographies/Lagrange/

Problem 4.1.5 Cauchy's Counterexample, Part 1.

(a) Adopting the notation $y = x^{-1}$ and $f^{(n)}(x) = p_n(y)e^{-y^2}$, find $p_{n+1}(y)$ in terms of $p_n(y)$. [Note: Don't forget that you are differentiating with respect to x, not y.]

(b) Use induction on n to show that $p_n(y)$ is a polynomial for all $n \in \mathbb{N}$.

□

Unfortunately everything we've done so far only gives us the derivatives we need when x is *not* zero, and we need the derivatives when x *is* zero. To find these we need to get back to very basic ideas.

Let's assume for the moment that we know that $f^{(n)}(0) = 0$ and recall that

$$f^{(n+1)}(0) = \lim_{x\to 0} \frac{f^{(n)}(x) - f^{(n)}(0)}{x - 0}$$
$$f^{(n+1)}(0) = \lim_{x\to 0} x^{-1} p_n(x^{-1})e^{-x^{-2}}$$
$$f^{(n+1)}(0) = \lim_{y\to\pm\infty} \frac{y p_n(y)}{e^{y^2}}.$$

We can close the deal with the following problem.

Problem 4.1.6 Cauchy's Counterexample, Part 2.

(a) Let m be a nonnegative integer. Show that $\lim_{y\to\pm\infty} \frac{y^m}{e^{y^2}} = 0$.

Hint.

Induction and a dash of L'Hôpital's rule should do the trick.

(b) Prove that $\lim_{y\to\pm\infty} \frac{q(y)}{e^{y^2}} = 0$ for any polynomial q.

(c) Let $f(x)$ be as in equation (I.1) and show that for every nonnegative integer n, $f^{(n)}(0) = 0$.

□

This example showed that while it was fruitful to exploit Taylor series representations of various functions, basing the foundations of calculus on power series was not a sound idea.

While Lagrange's approach wasn't totally successful, it was a major step away from infinitesimals and toward the modern approach. We still use aspects of it today. For instance we still use his prime notation (f') to denote the derivative.

Turning Lagrange's idea on its head it is clear that if we know how to compute derivatives, we can use this machine to obtain a power series when we are not "clever enough" to obtain the series in other (typically shorter) ways. For example, consider Newton's binomial series when $\alpha = \frac{1}{2}$. Originally, we obtained this series by extending the binomial theorem to non-integer exponents. Taylor's formula provides a more systematic way to obtain this series:

$$f(x) = (1+x)^{\frac{1}{2}}; \qquad f(0) = 1$$
$$f'(x) = \frac{1}{2}(1+x)^{\frac{1}{2}-1}; \qquad f'(0) = \frac{1}{2}$$
$$f''(x) = \frac{1}{2}\left(\frac{1}{2}-1\right)(1+x)^{\frac{1}{2}-2} \qquad f''(0) = \frac{1}{2}\left(\frac{1}{2}-1\right)$$

and in general since

$$f^{(n)}(x) = \frac{1}{2}\left(\frac{1}{2}-1\right)\cdots\left(\frac{1}{2}-(n-1)\right)(1+x)^{\frac{1}{2}-n}$$

we have

$$f^{(n)}(0) = \frac{1}{2}\left(\frac{1}{2}-1\right)\cdots\left(\frac{1}{2}-(n-1)\right).$$

Using Taylor's formula we obtain the series

$$\sum_{n=0}^{\infty}\frac{f^{(n)}(0)}{n!}x^n = 1+\sum_{n=1}^{\infty}\frac{\frac{1}{2}\left(\frac{1}{2}-1\right)\cdots\left(\frac{1}{2}-(n-1)\right)}{n!}x^n = 1+\sum_{n=1}^{\infty}\frac{\prod_{j=0}^{n-1}\left(\frac{1}{2}-j\right)}{n!}x^n$$

which agrees with equation (I.7) in the previous chapter.

Problem 4.1.7 Use Taylor's formula to obtain the general binomial series

$$(1+x)^{\alpha} = 1+\sum_{n=1}^{\infty}\frac{\prod_{j=0}^{n-1}(\alpha-j)}{n!}x^n.$$

□

Problem 4.1.8 Use Taylor's formula to obtain the Taylor series for the functions e^x, $\sin x$, and $\cos x$ expanded about a. □

As you can see, Taylor's "machine" will produce the power series for a function (if it has one), but is tedious to perform. We will find, generally, that this tediousness can be an obstacle to understanding. In many cases it will be better to be clever if we can. This is usually shorter. However, it is comforting to have Taylor's formula available as a last resort.

The existence of a Taylor series is addressed (to some degree) by the following.

Theorem 4.1.9 *If f', f'', $\ldots$, $f^{(n+1)}$ are all continuous on an interval containing a and x, then*

$$f(x) = f(a) + \frac{f'(a)}{1!}(x-a) + \frac{f''(a)}{2!}(x-a)^2 + \cdots + \frac{f^{(n)}(a)}{n!}(x-a)^n + \frac{1}{n!}\int_{t=a}^{x} f^{(n+1)}(t)(x-t)^n\,\mathrm{d}t.$$

Before we address the proof, notice that the n-th degree polynomial

$$f(a) + \frac{f'(a)}{1!}(x-a) + \frac{f''(a)}{2!}(x-a)^2 + \cdots + \frac{f^{(n)}(a)}{n!}(x-a)^n$$

resembles the Taylor series and, in fact, is called the *n-th degree Taylor polynomial of f about a.* Theorem 4.1.9 says that a function can be written as the sum of this polynomial and a specific integral which we will analyze in the next chapter. We will get the proof started and leave the formal induction proof as an exercise.

Notice that the case when $n = 0$ is really a restatement of the Fundamental Theorem of Calculus. Specifically, the FTC says $\int_{t=a}^{x} f'(t)\,\mathrm{d}t = f(x) - f(a)$ which we can rewrite as

$$f(x) = f(a) + \frac{1}{0!}\int_{t=a}^{x} f'(t)(x-t)^0\,\mathrm{d}t$$

to provide the anchor step for our induction.

To derive the case where $n = 1$, we use integration by parts. If we let

$$u = f'(t) \qquad dv = (x-t)^0 dt$$
$$du = f''(t)dt \qquad v = -\frac{1}{1}(x-t)^1$$

we obtain

$$\begin{aligned} f(x) &= f(a) + \frac{1}{0!}\left(-\frac{1}{1}f'(t)(x-t)^1\Big|_{t=a}^{x} + \frac{1}{1}\int_{t=a}^{x} f''(t)(x-t)^1\,\mathrm{dt}\right) \\ &= f(a) + \frac{1}{0!}\left(-\frac{1}{1}f'(x)(x-x)^1 + \frac{1}{1}f'(a)(x-a)^1 + \frac{1}{1}\int_{t=a}^{x} f''(t)(x-t)^1\,\mathrm{dt}\right) \\ &= f(a) + \frac{1}{1!}f'(a)\,(x-a)^1 + \frac{1}{1!}\int_{t=a}^{x} f''(t)(x-t)^1\,\mathrm{dt}. \end{aligned}$$

Problem 4.1.10 Provide a formal induction proof for Theorem 4.1.9. □

4.2 Series Anomalies

Up to this point, we have been somewhat frivolous in our approach to series. This approach mirrors eighteenth century mathematicians who ingeniously exploited calculus and series to provide mathematical and physical results which were virtually unobtainable before. Mathematicans were eager to push these techniques as far as they could to obtain their results and they often showed good intuition regarding what was mathematically acceptable and what was not. However, as the envelope was pushed, questions about the validity of the methods surfaced.

As an illustration consider the series expansion

$$\frac{1}{1+x} = 1 - x + x^2 - x^3 + \cdots.$$

If we substitute $x = 1$ into this equation, we obtain

$$\frac{1}{2} = 1 - 1 + 1 - 1 + \cdots.$$

If we group the terms as follows $(1-1)+(1-1)+\cdots$, the series would equal 0. A regrouping of $1 + (-1+1) + (-1+1) + \cdots$ provides an answer of 1. This violation of the associative law of addition did not escape the mathematicians of the 1700's. In his 1760 paper *On Divergent Series* Euler said:

> Notable enough, however are the controversies over the series $1 - 1 + 1 - 1 +$ *etc*, whose sum was given by Leibniz as $\frac{1}{2}$, although others disagree . . . Understanding of this question is to be sought in the word "sum;" this idea, if thus conceived - namely, the sum of a series is said to be that quantity to which it is brought closer as more terms of a series are taken - has relevance only for the convergent series, and we should in general give up this idea of sum for divergent series. On the other hand, as series in analysis arise from the expansion of fractions or irrational quantities or even of transcendentals, it will, in turn, be permissible in calculation to substitute in place of such series that quantity out of whose development it is produced.

Even with this formal approach to series, an interesting question arises. The series for the antiderivative of $\frac{1}{1+x}$ *does* converge for $x = 1$ while this one does not. Specifically, taking the antiderivative of the above series, we obtain

$$\ln(1 + x) = x - \frac{1}{2}x^2 + \frac{1}{3}x^3 - \cdots.$$

If we substitute $x = 1$ into this series, we obtain $\ln 2 = 1 - \frac{1}{2} + \frac{1}{3} - \cdots$. It is not hard to see that such an alternating series converges. The following picture shows why. In this diagram, S_n denotes the partial sum $1-\frac{1}{2}+\frac{1}{3}-\cdots+\frac{(-1)^{n+1}}{n}$.

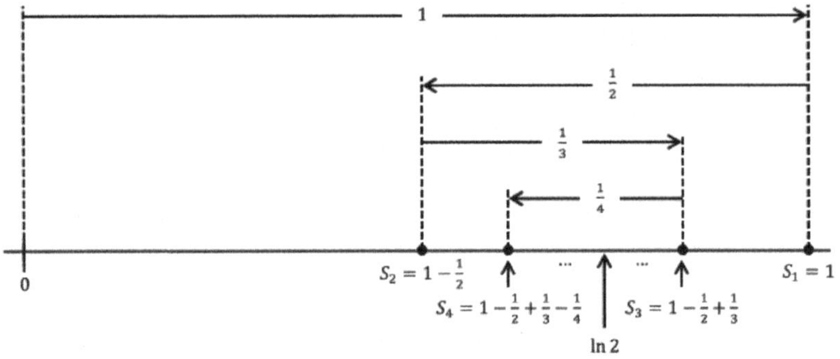

From the diagram we can see $S_2 \le S_4 \le S_6 \le \cdots \le \cdots \le S_5 \le S_3 \le S_1$ and $S_{2k+1} - S_{2k} = \frac{1}{2k+1}$. It seems that the sequence of partial sums will converge to whatever is in the "middle." Our diagram indicates that it is ln 2 in the middle but actually this is not obvious. Nonetheless it is interesting that one series converges for $x = 1$ but the other does not.

Problem 4.2.1 Use the fact that

$$1-\frac{1}{2}+\frac{1}{3}-\cdots+\frac{(-1)^{2k+1}}{2k} \le \ln 2 \le 1-\frac{1}{2}+\frac{1}{3}-\cdots+\frac{(-1)^{2k+2}}{2k+1}$$

to determine how many terms of the series $\sum_{n=1}^{\infty}\frac{(-1)^{n+1}}{n}$ should be added together to approximate $\ln 2$ to within .0001 without actually computing what $\ln 2$ is. □

There is an even more perplexing situation brought about by these examples. An infinite sum such as $1 - 1 + 1 - 1 + \cdots$ appears to not satisfy the associative law for addition. While a convergent series such as $1 - \frac{1}{2} + \frac{1}{3} - \cdots$ does satisfy the associative law, it does not satisfy the commutative law. In fact, it does not satisfy it rather spectacularly.

A generalization of the following result was stated and proved by Bernhard Riemann in 1854.

Theorem 4.2.2 *Let a be any real number. There exists a rearrangement of the series $1 - \frac{1}{2} + \frac{1}{3} - \cdots$ which converges to a.*

This theorem shows that a series is most decidedly *not* a great big sum. It follows that a power series is *not* a great big polynomial.

To set the stage, consider the *harmonic series*

$$\sum_{n=1}^{\infty}\frac{1}{n} = 1 + \frac{1}{2} + \frac{1}{3} + \cdots.$$

Even though the individual terms in this series converge to 0, the series still diverges (to infinity) as evidenced by the inequality

$$\left(1+\frac{1}{2}\right)+\left(\frac{1}{3}+\frac{1}{4}\right)+\left(\frac{1}{5}+\frac{1}{6}+\frac{1}{7}+\frac{1}{8}\right)+\left(\frac{1}{9}+\cdots+\frac{1}{16}\right)+\cdots$$

$$
\begin{aligned}
&> \frac{1}{2}+\left(\frac{1}{4}+\frac{1}{4}\right)+\left(\frac{1}{8}+\frac{1}{8}+\frac{1}{8}+\frac{1}{8}\right)+\left(\frac{1}{16}+\cdots+\frac{1}{16}\right)+\cdots \\
&= \frac{1}{2}+\frac{1}{2}+\frac{1}{2}+\frac{1}{2}+\cdots \\
&= \infty.
\end{aligned}
$$

Armed with this fact, we can see why Theorem 4.2.2 is true. First note that

$$-\frac{1}{2}-\frac{1}{4}-\frac{1}{6}-\cdots=-\frac{1}{2}(1+\frac{1}{2}+\frac{1}{3}+\cdots)=-\infty$$

and

$$1+\frac{1}{3}+\frac{1}{5}+\cdots\geq\frac{1}{2}+\frac{1}{4}+\frac{1}{6}+\ldots=\infty.$$

This says that if we add enough terms of $-\frac{1}{2}-\frac{1}{4}-\frac{1}{6}-\cdots$ we can make such a sum as small as we wish and if we add enough terms of $1+\frac{1}{3}+\frac{1}{5}+\cdots$ we can make such a sum as large as we wish. This provides us with the general outline of the proof. The trick is to add just enough positive terms until the sum is just greater than a. Then we start to add on negative terms until the sum is just less than a. Picking up where we left off with the positive terms, we add on just enough positive terms until we are just above a again. We then add on negative terms until we are below a. In essence, we are bouncing back and forth around a. If we do this carefully, then we can get this rearrangement to converge to a. The notation in the proof below gets a bit hairy, but keep this general idea in mind as you read through it.

Let O_1 be the first odd integer such that $1+\frac{1}{3}+\frac{1}{5}+\cdots+\frac{1}{O_1}>a$. Now choose E_1 to be the first even integer such that

$$-\frac{1}{2}-\frac{1}{4}-\frac{1}{6}-\cdots-\frac{1}{E_1}<a-\left(1+\frac{1}{3}+\frac{1}{5}+\cdots+\frac{1}{O_1}\right).$$

Thus

$$1+\frac{1}{3}+\frac{1}{5}+\cdots+\frac{1}{O_1}-\frac{1}{2}-\frac{1}{4}-\frac{1}{6}-\cdots-\frac{1}{E_1}<a.$$

Notice that we still have $\frac{1}{O_1+2}+\frac{1}{O_1+4}+\cdots=\infty$. With this in mind, choose O_2 to be the first odd integer with

$$\frac{1}{O_1+2}+\frac{1}{O_1+4}+\cdots\frac{1}{O_2}>a-\left(1+\frac{1}{3}+\frac{1}{5}+\cdots+\frac{1}{O_1}-\frac{1}{2}-\frac{1}{4}-\frac{1}{6}-\cdots-\frac{1}{E_1}\right).$$

Thus we have

$$a<1+\frac{1}{3}+\frac{1}{5}+\cdots+\frac{1}{O_1}-\frac{1}{2}-\frac{1}{4}-\frac{1}{6}-\cdots-\frac{1}{E_1}+\frac{1}{O_1+2}+\frac{1}{O_1+4}+\cdots+\frac{1}{O_2}.$$

Furthermore, since

$$1+\frac{1}{3}+\frac{1}{5}+\cdots+\frac{1}{O_1}-\frac{1}{2}-\frac{1}{4}-\frac{1}{6}-\cdots-\frac{1}{E_1}+\frac{1}{O_1+2}+\frac{1}{O_1+4}+\cdots+\frac{1}{O_2-2}<a$$

we have

$$
\begin{aligned}
&\left|1+\frac{1}{3}+\frac{1}{5}+\cdots+\frac{1}{O_1}-\frac{1}{2}-\frac{1}{4}-\frac{1}{6}-\cdots\right. \\
&\left.-\frac{1}{E_1}+\frac{1}{O_1+2}+\frac{1}{O_1+4}+\cdots+\frac{1}{O_2}-a\right|<\frac{1}{O_2}.
\end{aligned}
$$

In a similar fashion choose E_2 to be the first even integer such that

$$\begin{aligned} 1+\frac{1}{3}+\frac{1}{5}+\cdots+\frac{1}{O_1}-\frac{1}{2}-\frac{1}{4}-\frac{1}{6}- & \cdots \\ -\frac{1}{E_1}+\frac{1}{O_1+2}+\frac{1}{O_1+4}+\cdots \\ +\frac{1}{O_2}-\frac{1}{E_1+2}-\frac{1}{E_1+4}-\cdots \\ -\frac{1}{E_2}<a. \end{aligned}$$

Since

$$\begin{aligned} 1+\frac{1}{3}+\frac{1}{5}+\cdots+\frac{1}{O_1}-\frac{1}{2}-\frac{1}{4}-\frac{1}{6}-\cdots-\frac{1}{E_1} \\ +\frac{1}{O_1+2}+\frac{1}{O_1+4}+\cdots+\frac{1}{O_2}-\frac{1}{E_1+2}-\frac{1}{E_1+4}-\cdots-\frac{1}{E_2-2}>a \end{aligned}$$

then

$$\begin{aligned} \Bigg| 1+\frac{1}{3}+\frac{1}{5}+\cdots+\frac{1}{O_1}-\frac{1}{2}-\frac{1}{4}-\frac{1}{6}-\cdots-\frac{1}{E_1} \\ +\frac{1}{O_1+2}+\frac{1}{O_1+4}+\cdots+\frac{1}{O_2}-\frac{1}{E_1+2}-\frac{1}{E_1+4}-\cdots-\frac{1}{E_2}-a\Bigg| \\ <\frac{1}{E_2}. \end{aligned}$$

Again choose O_3 to be the first odd integer such that

$$\begin{aligned} a<1+\frac{1}{3}+\frac{1}{5}+\cdots+\frac{1}{O_1}-\frac{1}{2}-\frac{1}{4}-\frac{1}{6}-\cdots \\ -\frac{1}{E_1}+\frac{1}{O_1+2}+\frac{1}{O_1+4}+\cdots+\frac{1}{O_2}+\cdots \\ -\frac{1}{E_1+2}-\frac{1}{E_1+4}-\cdots-\frac{1}{E_2}+\frac{1}{O_2+2}+\frac{1}{O_2+4}+\cdots+\frac{1}{O_3} \end{aligned}$$

and notice that

$$\begin{aligned} \Bigg| 1+\frac{1}{3}+\frac{1}{5}+\cdots+\frac{1}{O_1}-\frac{1}{2}-\frac{1}{4} \\ -\frac{1}{6}-\cdots-\frac{1}{E_1}+\frac{1}{O_1+2}+\frac{1}{O_1+4}+\cdots+\frac{1}{O_2}+\cdots \\ -\frac{1}{E_1+2}-\frac{1}{E_1+4}-\cdots-\frac{1}{E_2}+\frac{1}{O_2+2}+\frac{1}{O_2+4}+\cdots+\frac{1}{O_3}-a\Bigg| \\ <\frac{1}{O_3}. \end{aligned}$$

Continue defining O_k and E_k in this fashion. Since $\lim_{k\to\infty}\frac{1}{O_k}=\lim_{k\to\infty}\frac{1}{E_k}=0$, it is evident that the partial sums

$$\begin{aligned} 1+\frac{1}{3}+\frac{1}{5}+\cdots+\frac{1}{O_1}-\frac{1}{2}-\frac{1}{4} \\ -\frac{1}{6}-\cdots-\frac{1}{E_1}+\frac{1}{O_1+2}+\frac{1}{O_1+4}+\cdots \\ +\frac{1}{O_2}+\cdots-\frac{1}{E_{k-2}+2}-\frac{1}{E_{k-2}+4}-\cdots \end{aligned}$$

$$-\frac{1}{E_{k-1}}+\frac{1}{O_{k-1}+2}+\frac{1}{O_{k-1}+4}+\cdots+\frac{1}{O_k}$$

and

$$1+\frac{1}{3}+\frac{1}{5}+\cdots+\frac{1}{O_1}-\frac{1}{2}-\frac{1}{4}-\frac{1}{6}$$
$$-\cdots-\frac{1}{E_1}+\frac{1}{O_1+2}+\frac{1}{O_1+4}+\cdots+\frac{1}{O_2}+\cdots$$
$$-\frac{1}{E_{k-2}+2}-\frac{1}{E_{k-2}+4}-\cdots-\frac{1}{E_{k-1}}$$

must converge to a. Furthermore, it is evident that every partial sum of the rearrangement

$$1+\frac{1}{3}+\frac{1}{5}+\cdots+\frac{1}{O_1}-\frac{1}{2}-\frac{1}{4}-\frac{1}{6}$$
$$-\cdots-\frac{1}{E_1}+\frac{1}{O_1+2}+\frac{1}{O_1+4}+\cdots+\frac{1}{O_2}+\cdots$$

is trapped between two such extreme partial sums. This forces the entire rearranged series to converge to a.

The next two problems are similar to the above, but notationally are easier since we don't need to worry about converging to an actual number. We only need to make the rearrangement grow (or shrink in the case of problem 4.2.4) without bound.

Problem 4.2.3 Show that there is a rearrangement of $1-\frac{1}{2}+\frac{1}{3}-\frac{1}{4}+\cdots$ which diverges to ∞. □

Problem 4.2.4 Show that there is a rearrangement of $1-\frac{1}{2}+\frac{1}{3}-\frac{1}{4}+\cdots$ which diverges to $-\infty$. □

It is fun to know that we can rearrange some series to make them add up to anything you like but there is a more fundamental idea at play here. That the negative terms of the alternating Harmonic Series *diverge* to negative infinity and the positive terms *diverge* to positive infinity make the *convergence* of the alternating series very special.

Consider, first we add 1. This is one of the positive terms so our sum is starting to increase without bound. Next we add $-1/2$ which is one of the negative terms so our sum has turned around and is now starting to decrease without bound. Then another positive term is added: increasing without bound. Then another negative term: decreasing. And so on. The convergence of the alternating Harmonic Series is the result of a delicate balance between a tendency to run off to positive infinity and back to negative infinity. When viewed in this light it is not really too surprising that rearranging the terms can destroy this delicate balance.

Naturally, the alternating Harmonic Series is not the only such series. Any such series is said to converge "conditionally" — the condition being the specific arrangement of the terms.

To stir the pot a bit more, some series do satisfy the commutative property. More specifically, one can show that any rearrangement of the series $1-\frac{1}{2^2}+\frac{1}{3^2}-\cdots$ must converge to the same value as the original series (which happens to be $\int_{x=0}^{1}\frac{\ln(1+x)}{x}dx\approx .8224670334$). Why does one series behave so nicely whereas the other does not?

Issues such as these and, more generally, the validity of using the infinitely small and infinitely large certainly existed in the 1700's, but they were overshadowed by the utility of the calculus. Indeed, foundational questions raised by the above examples, while certainly interesting and of importance, did not significantly deter the exploitation of calculus in studying physical phenomena. However, the envelope eventually was pushed to the point that not even the most practically oriented mathematician could avoid the foundational issues.

4.3 Additional Problems

Problem 4.3.1 Use Taylor's formula to find the Taylor series of the given function expanded about the given point a.

(a) $f(x) = \ln(1 + x)$, $a = 0$

(b) $f(x) = e^x$, $a = -1$

(c) $f(x) = x^3 + x^2 + x + 1$, $a = 0$

(d) $f(x) = x^3 + x^2 + x + 1$, $a = 1$

□

Part II

Interregnum

Chapter 5

Joseph Fourier: The Man Who Broke Calculus

5.1 Joseph Fourier and His Series

Applying mathematics to physical problems such as heat flow in a solid body drew much attention in the latter part of the 1700's and the early part of the 1800's. One of the people to attack the heat flow problem was

Figure 5.1.1 Jean Baptiste Joseph Fourier[13]

Jean Baptiste Joseph Fourier. Fourier submitted a manuscript on the subject, *Sur la propagation de la chaleur* (*On the Propagation of Heat*), to the *Institut National des Sciences et des Arts* in 1807. These ideas were subsequently published in *La theorie analytique de la chaleur* (*The Analytic Theory of Heat (1822)*).

To examine Fourier's ideas, consider the example of a thin wire of length one, which is perfectly insulated and whose endpoints are held at a fixed temperature of zero. Given an initial temperature distribution in the wire, the problem is to monitor the temperature of the wire at any point x and at any time t. Specifically, if we let $u(x,t)$ denote the temperature of the wire at point $x \in [0,1]$ at time $t \geq 0$, then it can be shown that u must satisfy the one-dimensional heat equation $\rho^2 \frac{\partial^2 u}{\partial x^2} = \frac{\partial u}{\partial t}$, where ρ^2 is a positive constant known as the thermal diffusivity. If the initial temperature distribution is given by

[13] `mathshistory.st-andrews.ac.uk/Biographies/Fourier/`

the function $f(x)$, then the u we are seeking must satisfy all of the following

$$\rho^2 \frac{\partial^2 u}{\partial x^2} = \frac{\partial u}{\partial t}$$

$$u(0,t) = u(1,t) = 0, \quad \forall t \geq 0$$
$$u(x,0) = f(x), \quad \forall x \in [0,1].$$

To solve this, Fourier employed what is now referred to as Fourier's method of separation of variables. Specifically, Fourier looked for solutions of the form $u(x,t) = X(x)T(t)$; that is, solutions where the x-part can be separated from the t-part. Assuming that u has this form, we get $\frac{\partial^2 u}{\partial x^2} = X''T$ and $\frac{\partial u}{\partial t} = XT'$. Substituting these into the differential equation $\rho^2 \frac{\partial^2 u}{\partial x^2} = \frac{\partial u}{\partial t}$, we obtain

$$\rho^2 X''T = XT' \text{ or } \frac{X''}{X} = \frac{T'}{\rho^2 T}.$$

Since the left-hand side involves no t's and the right-hand side involves no x's, both sides must equal a constant k. Thus we have

$$X'' = kX \text{ and } T' = \rho^2 kT.$$

Problem 5.1.2 Show that $T = Ce^{\rho^2 kt}$ satisfies the equation $T' = \rho^2 kT$, where C, and ρ are arbitrary constants. Use the physics of the problem to show that if u is not constantly zero, then $k < 0$.

Hint.

Consider $\lim_{t \to \infty} u(x,t)$. □

Using the result from problem 5.1.2 that $k < 0$, we will let $k = -p^2$.

Problem 5.1.3 Show that $X = A\sin(px) + B\cos(px)$ satisfies the equation $X'' = -p^2 X$, where A and B are arbitrary constants. Use the boundary conditions $u(0,t) = u(1,t) = 0$, $\forall t \geq 0$ to show that $B = 0$ and $A \sin p = 0$. Conclude that if u is not constantly zero, then $p = n\pi$, where n is any integer. □

Problem 5.1.4

Show that if u_1 and u_2 satisfy the equations $\rho^2 \frac{\partial^2 u}{\partial x^2} = \frac{\partial u}{\partial t}$ and $u(0,t) = u(1,t) = 0, \forall t \geq 0$ then $u = A_1 u_1 + A_2 u_2$ satisfy these as well, where A_1 and A_2 are arbitrary constants. □

Putting all of these results together, Fourier surmised that the general solution to

$$\rho^2 \frac{\partial^2 u}{\partial x^2} = \frac{\partial u}{\partial t} u(0,t) = u(1,t) = 0, \forall t \geq 0$$

could be expressed as the series

$$u(x,t) = \sum_{n=1}^{\infty} A_n e^{-(\rho n\pi)^2 t} \sin(n\pi x).$$

All that is left is to have u satisfy the initial condition $u(x,0) = f(x)$, $\forall x \in [0,1]$. That is, we need to find coefficients A_n, such that

$$f(x) = u(x,0) = \sum_{n=1}^{\infty} A_n \sin(n\pi x).$$

The idea of representing a function as a series of sine waves was proposed by Daniel Bernoulli in 1753 while examining the problem of modeling a vibrating string. Unfortunately for Bernoulli, he didn't know how to compute the coefficients in such a series representation. What distinguished Fourier was that he developed a technique to compute these coefficients. The key is the result of the following problem.

Problem 5.1.5 Let n and m be positive integers. Show

$$\int_{x=0}^{1} \sin(n\pi x)\sin(m\pi x)\,dx = \begin{cases} 0 & \text{if } n \neq m \\ \frac{1}{2} & \text{if } n = m \end{cases}.$$

□

Armed with the result from Problem 5.1.5, Fourier could compute the coefficients A_n in the series representation $f(x) = \sum_{n=1}^{\infty} A_n \sin(n\pi x)$ in the following manner. Since we are trying to find A_n for a particular (albeit general) n, we will temporarily change the index in the summation from n to j. With this in mind, consider

$$\begin{aligned}\int_{x=0}^{1} f(x)\sin(n\pi x)\,dx &= \int_{x=0}^{1}\left(\sum_{j=1}^{\infty} A_j \sin(j\pi x)\right)\sin(n\pi x)\,dx \\ &= \sum_{j=1}^{\infty} A_j \int_{x=0}^{1} \sin(j\pi x)\sin(n\pi x)\,dx \\ &= A_n \cdot \frac{1}{2}\end{aligned}$$

This leads to the formula $A_n = 2\int_{x=0}^{1} f(x)\sin(n\pi x)\,dx$.

The above series $f(x) = \sum_{n=1}^{\infty} A_n \sin(n\pi x)$ with $A_n = 2\int_{x=0}^{1} f(x)\sin(n\pi x)\,dx$ is called the *Fourier (sine) series of* $\boldsymbol{f}$.

Example 5.1.6 Let's apply this to the following function, $f(x) = \frac{1}{2} - \left|x - \frac{1}{2}\right|$, whose graph of this function is seen below.

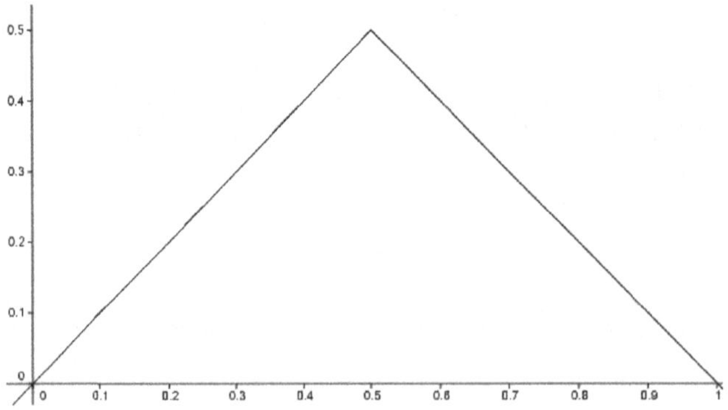

Graph of $f(x) = \frac{1}{2} - \left|x - \frac{1}{2}\right|$

□

Problem 5.1.7 Let n be a positive integer. Show that if

$$f(x) = \frac{1}{2} - \left|x - \frac{1}{2}\right|$$

then

$$\int_{x=0}^{1} f(x)\sin(n\pi x)\,dx = \frac{2}{(n\pi)^2}\sin\left(\frac{n\pi}{2}\right)$$

and show that the Fourier sine series of f is given by

$$f(x) = \sum_{n=1}^{\infty} \frac{4}{(n\pi)^2} \sin\left(\frac{n\pi}{2}\right) \sin(n\pi x) = \frac{4}{\pi^2} \sum_{k=0}^{\infty} \frac{(-1)^k}{(2k+1)^2} \sin((2k+1)\pi x).$$

□

To check if this series really does, in fact, represent f on $[0, 1]$, let

$$S_N(x) = \frac{4}{\pi^2} \sum_{k=0}^{N} \frac{(-1)^k}{(2k+1)^2} \sin((2k+1)\pi x)$$

be the Nth partial sum of the series and use the graphing tool below to view the graph of $S_N(x)$ for several values of N.

That is, S_N denotes the Nth partial sum of the series. We will graph S_N for $N = 1, 2, 5, 50$.

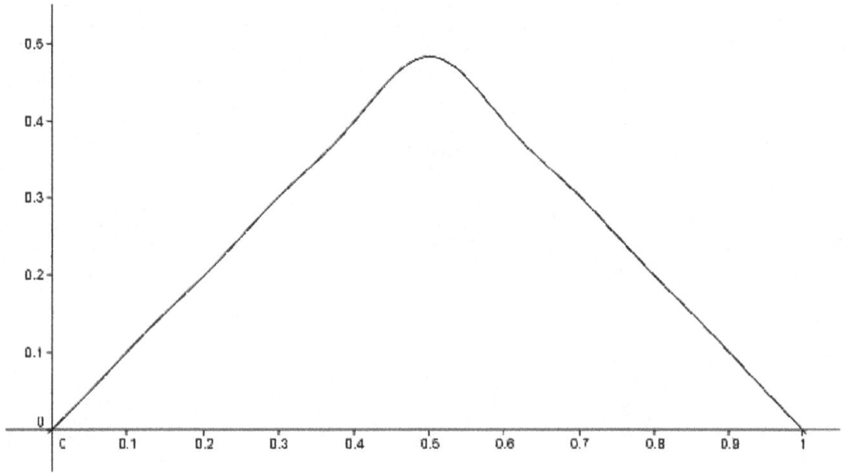

$$S_5(x) = \frac{4}{\pi^2} \sum_{k=0}^{5} \frac{(-1)^k}{(2k+1)^2} \sin((2k+1)\pi x)$$

As you can see, it appears that as we add more terms to the partial sum, S_N, it looks more and more like the original function $f(x) = \frac{1}{2} - \left|x - \frac{1}{2}\right|$. This would lead us to believe that the series converges to the function and that

$$f(x) = \frac{4}{\pi^2} \sum_{k=0}^{\infty} \frac{(-1)^k}{(2k+1)^2} \sin((2k+1)\pi x). \tag{II.1}$$

is a valid representation of f as a Fourier series.

Recall, that when we represented a function as a power series, we freely differentiated and integrated the series term by term as though it was a polynomial. Let's do the same with this Fourier series.

To start, notice that the derivative of

$$f(x) = \frac{1}{2} - \left|x - \frac{1}{2}\right|$$

is given by

$$f'(x) = \begin{cases} 1 & \text{if } 0 \le x < \frac{1}{2} \\ -1 & \text{if } \frac{1}{2} < x \le 1 \end{cases}.$$

This derivative does not exist at $x = \frac{1}{2}$ and its graph is given by

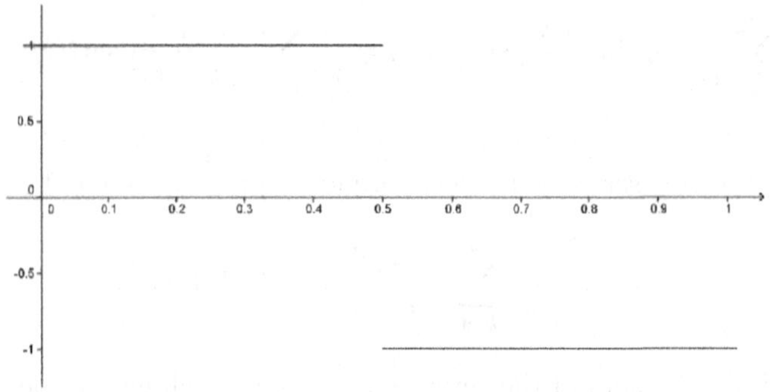

Graph of $f'(x)$

If we differentiate the Fourier series term-by-term, we obtain

$$\frac{4}{\pi}\sum_{k=0}^{\infty}\frac{(-1)^k}{(2k+1)}\cos\left((2k+1)\,\pi x\right).$$

Again, if we let $C_N(x) = \frac{4}{\pi}\sum_{k=0}^{N}\frac{(-1)^k}{(2k+1)}\cos\left((2k+1)\,\pi x\right)$ be the Nth partial sum of this Fourier cosine series and plot $C_N(x)$ for $N = 5$, we obtain

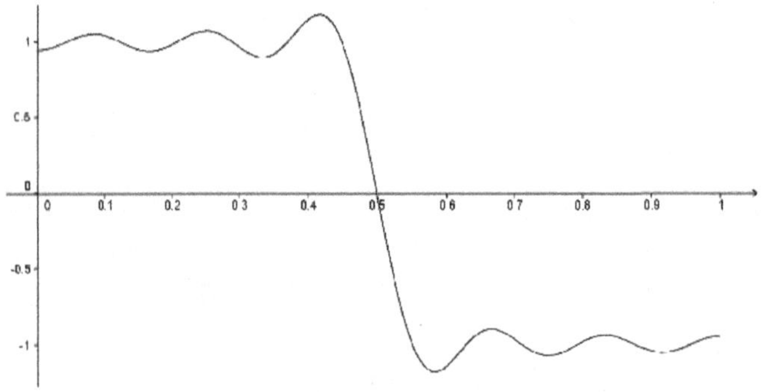

$$C_5(x) = \frac{4}{\pi}\sum_{k=0}^{5}\frac{(-1)^k}{2k+1}\cos((2k+1)\pi x)$$

In fact, if we were to graph the series $\frac{4}{\pi}\sum_{k=0}^{\infty}\frac{(-1)^k}{(2k+1)}\cos((2k+1)\,\pi x)$, we would obtain

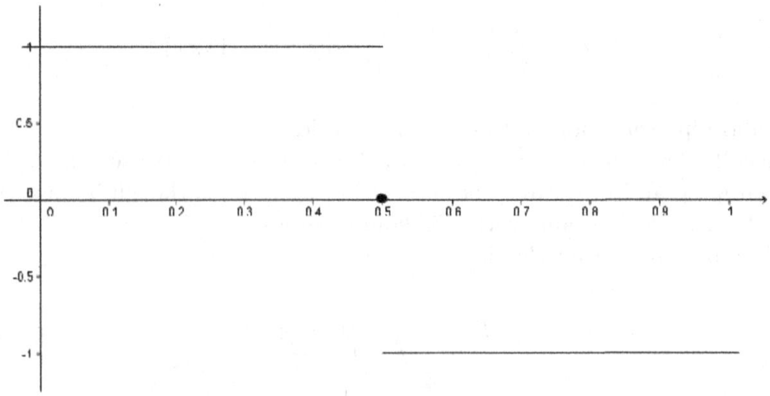

Graph of $\frac{4}{\pi}\sum_{k=0}^{\infty}\frac{(-1)^k}{2k+1}\cos((2k+1)\pi x)$

Notice that this agrees with the graph of f', except that f' didn't exist at $x = \frac{1}{2}$, and this series takes on the value 0 at $x = \frac{1}{2}$. Notice also, that every partial sum of this series is continuous, since it is a finite combination of continuous cosine functions. This agrees with what you learned in calculus, the (finite) sum of continuous functions is always continuous. In the 1700's, this was also assumed to be true for infinite series, because every time a power series converged to a function, that function happened to be continuous. This never failed for power series, so this example was a bit disconcerting as it is an example of the sum of *infinitely many* continuous functions which is, in this case, discontinuous. Was it possible that there was some power series which converged to a function which was not continuous? Even if there wasn't, what was the difference between power series and this Fourier series?

Even more disconcerting is what happens if we try differentiating the series

$$\frac{4}{\pi}\sum_{k=0}^{\infty}\frac{(-1)^k}{(2k+1)}\cos\left((2k+1)\,\pi x\right)$$

term-by-term. Given the above graph of this series, the derivative of it should be constantly 0, except at $x = \frac{1}{2}$, where the derivative wouldn't exist. Using the old adage that the derivative of a sum is the sum of the derivatives, we differentiate this series term-by-term to obtain the series

$$4\sum_{k=0}^{\infty}(-1)^{k+1}\sin\left((2k+1)\,\pi x\right).$$

If we sum the first forty terms of this series, we get

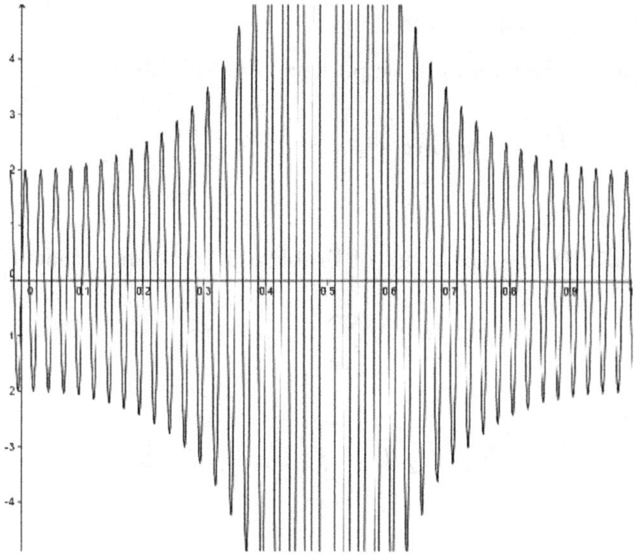

Graph of $4\sum_{k=0}^{39}(-1)^{k+1}\sin((2k+1)\pi x)$

We knew that there might be a problem at $x = \frac{1}{2}$ but this is crazy! The series seems to not be converging to zero at all!

Problem 5.1.8 Show that when $x = \frac{1}{4}$

$$4\sum_{k=0}^{\infty}(-1)^{k+1}\sin\left((2k+1)\,\pi x\right) = 4\left(-\frac{1}{\sqrt{2}}+\frac{1}{\sqrt{2}}+\frac{1}{\sqrt{2}}-\frac{1}{\sqrt{2}}-\frac{1}{\sqrt{2}}+\cdots\right).$$

□

Problem 5.1.8 shows that when we differentiate the series

$$\frac{4}{\pi}\sum_{k=0}^{\infty}\frac{(-1)^k}{(2k+1)}\cos\left((2k+1)\pi x\right)$$

term by term, this differentiated series doesn't converge to anything at $x=\frac{1}{4}$, let alone converge to zero. In this case, the old calculus rule that the derivative of a sum is the sum of the derivatives does not apply for this infinite sum, though it did apply before. As if the continuity issue wasn't bad enough before, this was even worse. Power series were routinely differentiated and integrated term-by-term. This was part of their appeal. They were treated like "infinite polynomials." Either there is some power series lurking that refuses to behave nicely, or there is some property that power series have that not all Fourier series have.

Could it be that everything we did in Chapter 4 was bogus?

Fortunately, the answer to that question is no. Power series are generally much more well-behaved than Fourier series. Whenever a power series converges, the function it converges to will be continuous. As long as one stays inside the interval of convergence, power series can be differentiated and integrated term-by-term. Power series have something going for them that your average Fourier series does not. (We need to develop the machinery to know what that something is.) None of this is any more obvious to us than it was to mathematicians at the beginning of the nineteenth century. What they did know was that relying on intuition was perilous and rigorous formulations were needed to either justify or dismiss these intuitions. In some sense, the nineteenth century was the "morning after" the mathematical party that went on throughout the eighteenth century.

Problem 5.1.9 Let n and m be positive integers. Show

$$\int_{x=0}^{1}\cos(n\pi x)\cos(m\pi x)\,\mathrm{d}x=\begin{cases}0 & \text{if } n\neq m\\ \frac{1}{2} & \text{if } n=m\end{cases}.$$

□

Problem 5.1.10 Use the result of Problem 5.1.9 to show that if

$$f(x)=\sum_{n=1}^{\infty}B_n\cos(n\pi x)$$

on $[0,1]$, then

$$B_m=2\int_{x=0}^{1}f(x)\cos(m\pi x)\,\mathrm{d}x.$$

□

Problem 5.1.11 Apply the result of Problem 5.1.10 to show that the Fourier cosine series of $f(x)=x-\frac{1}{2}$ on $[0,1]$ is given by

$$\frac{-4}{\pi^2}\sum_{k=0}^{\infty}\frac{1}{(2k+1)^2}\cos\left((2k+1)\pi x\right).$$

Let $C(x,N)=\frac{-4}{\pi^2}\sum_{k=0}^{N}\frac{1}{(2k+1)^2}\cos\left((2k+1)\pi x\right)$ and plot $C(x,N)$ for $N=1,2,5,50$ $x\in[0,1]$. How does this compare to the function $f(x)=x-\frac{1}{2}$ on $[0,1]$? What if you plot it for $x\in[0,2]$? □

Problem 5.1.12

(a) Differentiate the series

$$\frac{-4}{\pi^2}\sum_{k=0}^{\infty}\frac{1}{(2k+1)^2}\cos\left((2k+1)\pi x\right)$$

term by term and plot various partial sums for that series on $[0,1]$. How does this compare to the derivative of $f(x) = x - \frac{1}{2}$ on that interval?

(b) Differentiate the series you obtained in part a and plot various partial sums of that on $[0,1]$. How does this compare to the second derivative of $f(x) = x - \frac{1}{2}$ on that interval?

□

Part III

In Which We Find (Some) Answers

Chapter 6

Convergence of Sequences and Series

6.1 Sequences of Real Numbers

In Chapter 3, we developed the equation $1 + x + x^2 + x^3 + \cdots = \frac{1}{1-x}$, and we mentioned there were limitations to this power series representation. For example, substituting $x = 1$ and $x = -1$ into this expression leads to

$$1 + 1 + 1 + \cdots = \frac{1}{0} \text{ and } 1 - 1 + 1 - 1 + \cdots = \frac{1}{2}$$

which are rather hard to accept. On the other hand, if we substitute $x = \frac{1}{2}$ into the expression we get $1 + \frac{1}{2} + \left(\frac{1}{2}\right)^2 + \left(\frac{1}{2}\right)^3 + \cdots = 2$ which seems more palatable until we think about it. We can add two numbers together by the method we all learned in elementary school. Or three. Or any finite set of numbers, at least in principle. But infinitely many? What does that even mean? Before we can add infinitely many numbers together we must find a way to give meaning to the idea.

To do this, we examine an infinite sum by thinking of it as a sequence of finite partial sums. In our example, we would have the following sequence of partial sums.

$$\left(1, 1 + \frac{1}{2}, 1 + \frac{1}{2} + \left(\frac{1}{2}\right)^2, 1 + \frac{1}{2} + \left(\frac{1}{2}\right)^3, \ldots, \sum_{j=0}^{n}\left(\frac{1}{2}\right)^j, \ldots\right).$$

We can plot these sums on a number line to see what they tend toward as n gets large.

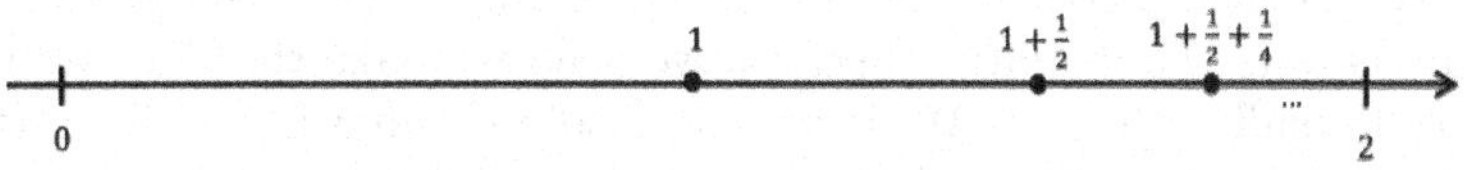

Since each partial sum is located at the midpoint between the previous partial sum and 2, it is reasonable to suppose that these sums tend to the number 2. Indeed, you probably have seen an expression such as $\lim_{n\to\infty}\left(\sum_{j=0}^{n}\left(\frac{1}{2}\right)^j\right) = 2$ justified by a similar argument. Of course, the reliance on such pictures and words is fine if we are satisfied with intuition. However, we must be able to make these intuitions rigorous without relying on pictures or nebulous words such as "approaches."

No doubt you are wondering "What's wrong with the word 'approaches'? It seems clear enough to me." This is often a sticking point. But if we think carefully about what we mean by the word "approach" we see that there is an implicit assumption that will cause us some difficulties later if we don't expose it.

To see this consider the sequence $\left(1, \frac{1}{2}, \frac{1}{3}, \frac{1}{4}, \ldots\right)$. Clearly it "approaches" zero, right? But, doesn't it also "approach" -1? It does, in the sense that each term gets closer to -1 than the one previous. But it also "approaches" -2, -3, or even -1000 in the same sense. That's the problem with the word "approaches." It just says that we're getting closer to something than we were in the previous step. It does *not* tell us that we are actually getting close. Since the moon moves in an elliptical orbit about the earth for part of each month it is "approaching" the earth. The moon gets clos*er* to the earth but, thankfully, it does not get *close* to the earth. The implicit assumption we alluded to earlier is this: When we say that the sequence $\left(\frac{1}{n}\right)_{n=1}^{\infty}$ "approaches" zero we mean that it is getting *close* not clos*er*. Ordinarily this kind of vagueness in our language is pretty innocuous. When we say "approaches" in casual conversation we can usually tell from the context of the conversation whether we mean "getting close to" or "getting closer to." But when speaking mathematically we need to be more careful, more explicit, in the language we use.

So how can we change the language we use so that this ambiguity is eliminated? Let's start out by recognizing, rigorously, what we mean when we say that a sequence converges to zero. For example, you would probably want to say that the sequence $\left(1, \frac{1}{2}, \frac{1}{3}, \frac{1}{4}, \ldots\right) = \left(\frac{1}{n}\right)_{n=1}^{\infty}$ converges to zero. Is there a way to give this meaning without relying on pictures or intuition?

One way would be to say that we can make $\frac{1}{n}$ as close to zero as we wish, provided we make n large enough. But even this needs to be made more specific. For example, we can get $\frac{1}{n}$ to within a distance of .1 of 0 provided we make $n > 10$, we can get $\frac{1}{n}$ to within a distance of .01 of 0 provided we make $n > 100$, etc. After a few such examples it is apparent that given any arbitrary distance $\varepsilon > 0$, we can get $\frac{1}{n}$ to within ε of 0 provided we make $n > \frac{1}{\varepsilon}$. This leads to the following definition.

Definition 6.1.1 Let $(s_n) = (s_1, s_2, s_3, \ldots)$ be a sequence of real numbers. We say that (s_n) **converges to 0** and write $\lim_{n\to\infty} s_n = 0$ provided for any $\varepsilon > 0$, there is a real number N such that if $n > N$, then $|s_n| < \varepsilon$. $\diamondsuit$

Notes on Definition 6.1.1:

1. This definition is the formal version of the idea we just talked about; that is, given an arbitrary distance ε, we must be able to find a specific number N such that s_n is within ε of 0, whenever $n > N$. The N is the answer to the question of how large is "large enough" to put s_n this close to 0.

2. Even though we didn't need it in the example $\left(\frac{1}{n}\right)$, the absolute value appears in the definition because we need to make the *distance* from s_n to 0 smaller than ε. Without the absolute value in the definition, we would be able to "prove" such outrageous statements as $\lim_{n\to\infty} -n = 0$, which we obviously don't want.

3. The statement $|s_n| < \varepsilon$ can also be written as $-\varepsilon < s_n < \varepsilon$ or $s_n \in (-\varepsilon, \varepsilon)$. (See the Problem 6.1.2 below.) Any one of these equivalent formulations can be used to prove convergence. Depending on the application, one of these may be more advantageous to use than the others.

4. Any time an N can be found that works for a particular ε, any number $M > N$ will work for that ε as well, since if $n > M$ then $n > N$.

Problem 6.1.2 Let a and b be real numbers with $b > 0$. Prove $|a| < b$ if and only if $-b < a < b$. Notice that this can be extended to $|a| \leq b$ if and only if $-b \leq a \leq b$. □

To illustrate how this definition makes the above ideas rigorous, let's use it to prove that $\lim_{n\to\infty} \frac{1}{n} = 0$.

Proof. Let $\varepsilon > 0$ be given. Let $N = \frac{1}{\varepsilon}$. If $n > N$, then $n > \frac{1}{\varepsilon}$ and so $|\frac{1}{n}| = \frac{1}{n} < \varepsilon$. Hence by definition, $\lim_{n\to\infty} \frac{1}{n} = 0$. ■

Notice that this proof is rigorous and makes no reference to vague notions such as "getting smaller" or "approaching infinity." It has three components:

1. provide the challenge of a distance $\varepsilon > 0$,
2. identify a real number N, and
3. show that this N works for this given ε.

There is also no explanation about where N came from. While it is true that this choice of N is not surprising in light of the "scrapwork" we did before the definition, the motivation for how we got it is not in the formal proof nor is it required. In fact, such scrapwork is typically not included in a formal proof. For example, consider the following.

Example 6.1.3 Use the definition of convergence to zero to prove

$$\lim_{n\to\infty} \frac{\sin n}{n} = 0.$$

□

Proof. Let $\varepsilon > 0$. Let $N = \frac{1}{\varepsilon}$. If $n > N$, then $n > \frac{1}{\varepsilon}$ and $\frac{1}{n} < \varepsilon$. Thus $\left|\frac{\sin(n)}{n}\right| \leq \frac{1}{n} < \varepsilon$. Hence by definition, $\lim_{n\to\infty} \frac{\sin n}{n} = 0$. ■

Notice that the N came out of nowhere, but you can probably see the thought process that went into this choice: We needed to use the inequality $|\sin n| \leq 1$. Again this scrapwork is not part of the formal proof, but it is typically necessary for finding what N should be. You might be able to do the next problem without doing any scrapwork first, but don't hesitate to do scrapwork if you need it.

Problem 6.1.4 Use the definition of convergence to zero to prove the following.

(a) $\lim_{n\to\infty} \frac{1}{n^2} = 0$

(b) $\lim_{n\to\infty} \frac{1}{\sqrt{n}} = 0$

□

As the sequences get more complicated, doing scrapwork ahead of time will become more necessary.

Example 6.1.5 Use the definition of convergence to zero to prove

$$\lim_{n\to\infty} \frac{n+4}{n^2+1} = 0.$$

SCRAPWORK

Given an $\varepsilon > 0$, we need to see how large to make n in order to guarantee that $|\frac{n+4}{n^2+1}| < \varepsilon$. First notice that $\frac{n+4}{n^2+1} < \frac{n+4}{n^2}$. Also, notice that if $n > 4$, then $n + 4 < n + n = 2n$. So as long as $n > 4$, we have $\frac{n+4}{n^2+1} < \frac{n+4}{n^2} < \frac{2n}{n^2} = \frac{2}{n}$. We can make this less than ε if we make $n > \frac{2}{\varepsilon}$. This means we need to make $n > 4$ and $n > \frac{2}{\varepsilon}$, simultaneously. These can be done if we let N be the maximum of these two numbers. This sort of thing comes up regularly, so the notation $N = \max\left(4, \frac{2}{\varepsilon}\right)$ was developed to mean the maximum of these two numbers. Notice that if $N = \max\left(4, \frac{2}{\varepsilon}\right)$ then $N \geq 4$ and $N \geq \frac{2}{\varepsilon}$. We're now ready for the formal proof. □

Proof. Let $\varepsilon > 0$. Let $N = \max\left(4, \frac{2}{\varepsilon}\right)$. If $n > N$, then $n > 4$ and $n > \frac{2}{\varepsilon}$. Thus we have $n > 4$ and $\frac{2}{n} < \varepsilon$. Therefore

$$\left|\frac{n+4}{n^2+1}\right| = \frac{n+4}{n^2+1} < \frac{n+4}{n^2} < \frac{2n}{n^2} = \frac{2}{n} < \varepsilon.$$

Hence by definition, $\lim_{n\to\infty} \frac{n+4}{n^2+1} = 0$. ■

Again we emphasize that the scrapwork is not ***explicitly*** a part of the formal proof. However, if you look carefully, you can always find the scrapwork in the formal proof.

Problem 6.1.6 Use the definition of convergence to zero to prove

$$\lim_{n\to\infty} \frac{n^2+4n+1}{n^3} = 0.$$

□

Problem 6.1.7 Let b be a nonzero real number with $|b| < 1$ and let $\varepsilon > 0$.

(a) Solve the inequality $|b|^n < \varepsilon$ for n

(b) Use part (a) to prove $\lim_{n\to\infty} b^n = 0$.

□

We can negate this definition to prove that a particular sequence does not converge to zero.

Example 6.1.8 Use the definition to prove that the sequence

$$(1 + (-1)^n)_{n=0}^{\infty} = (2, 0, 2, 0, 2, \ldots)$$

does not converge to zero. □

Before we provide this proof, let's analyze what it means for a sequence (s_n) to *not* converge to zero. Converging to zero means that any time a distance $\varepsilon > 0$ is given, we must be able to respond with a number N such that $|s_n| < \varepsilon$ for every $n > N$. To have this not happen, we must be able to find some $\varepsilon > 0$ such that no choice of N will work. Of course, if we find such an ε, then any smaller one will fail to have such an N, but we only need one to mess us up. If you stare at the example long enough, you see that any ε with $0 < \varepsilon \leq 2$ will cause problems. For our purposes, we will let $\varepsilon = 2$.

Proof. Let $\varepsilon = 2$ and let $N \in \mathbb{N}$ be any integer. If we let k be any non-negative integer with $k > \frac{N}{2}$, then $n = 2k > N$, but $|1 + (-1)^n| = 2$. Thus no choice of N will satisfy the conditions of the definition for this ε, (namely that $|1 + (-1)^n| < 2$ for all $n > N$) and so $\lim_{n\to\infty} (1 + (-1)^n) \neq 0$. ■

Problem 6.1.9 Negate the definition of $\lim_{n\to\infty} s_n = 0$ to provide a formal definition for $\lim_{n\to\infty} s_n \neq 0$. □

Problem 6.1.10 Use the definition to prove $\lim_{n\to\infty} \frac{n}{n+100} \neq 0$. □

Now that we have a handle on how to rigorously prove that a sequence converges to zero, let's generalize this to a formal definition for a sequence converging to something else. Basically, we want to say that a sequence (s_n) converges to a real number s, provided the difference $(s_n - s)$ converges to zero. This leads to the following definition:

Definition 6.1.11 Let $(s_n) = (s_1, s_2, s_3, \ldots)$ be a sequence of real numbers and let s be a real number. We say that (s_n) {converges to} s and write $\lim_{n\to\infty} s_n = s$ provided for any $\varepsilon > 0$, there is a real number N such that if $n > N$, then $|s_n - s| < \varepsilon$. ◊

{Notes on Definition 6.1.11}

1. Clearly
$$\lim_{n\to\infty} s_n = s \text{ if and only if } \lim_{n\to\infty} (s_n - s) = 0.$$

2. Again notice that this says that we can make s_n as close to s as we wish (within ε) by making n large enough ($> N$). As before, this definition makes these notions very specific.

3. Notice that $|s_n - s| < \varepsilon$ can be written in the following equivalent forms

   (a) $|s_n - s| < \varepsilon$

   (b) $-\varepsilon < s_n - s < \varepsilon$

   (c) $s - \varepsilon < s_n < s + \varepsilon$

   (d) $s_n \in (s - \varepsilon, s + \varepsilon)$

   and we are free to use any one of these which is convenient at the time.

As an example, let's use this definition to prove that the sequence in Problem 6.1.10, in fact, converges to 1.

Example 6.1.12 Prove $\displaystyle\lim_{n\to\infty} \frac{n}{n+100} = 1$.

SCRAPWORK

Given an $\varepsilon > 0$, we need to get $\left|\frac{n}{n+100} - 1\right| < \varepsilon$. This prompts us to do some algebra.
$$\left|\frac{n}{n+100} - 1\right| = \left|\frac{n-(n+100)}{n+100}\right| \leq \frac{100}{n}.$$

This in turn, seems to suggest that $N = \frac{100}{\varepsilon}$ should work. □

Proof. Let $\varepsilon > 0$. Let $N = \frac{100}{\varepsilon}$. If $n > N$, then $n > \frac{100}{\varepsilon}$ and so $\frac{100}{n} < \varepsilon$. Hence
$$\left|\frac{n}{n+100} - 1\right| = \left|\frac{n-(n+100)}{n+100}\right| = \frac{100}{n+100} < \frac{100}{n} < \varepsilon.$$

Thus by definition $\lim_{n\to\infty} \frac{n}{n+100} = 1$. ■

Notice again that the scrapwork is not part of the formal proof and the author of a proof is not obligated to tell where the choice of N came from (although the thought process can usually be seen in the formal proof). The formal proof contains only the requisite three parts: provide the challenge of an arbitrary $\varepsilon > 0$, provide a specific N, and show that this N works for the given ε.

Also notice that given a specific sequence such as $\left(\frac{n}{n+100}\right)$, the definition does not indicate what the limit would be if, in fact, it exists. Once an educated guess is made as to what the limit should be, the definition only verifies that this intuition is correct.

This leads to the following question: If intuition is needed to determine what a limit of a sequence should be, then what is the purpose of this relatively non-intuitive, complicated definition?

Remember that when these rigorous formulations were developed, intuitive notions of convergence were already in place and had been used with great success. This definition was developed to address the foundational issues. Could our intuitions be verified in a concrete fashion that was above reproach? This was the purpose of this non-intuitive definition. It was to be used to verify that our intuition was, in fact, correct and do so in a very prescribed manner. For example, if $b > 0$ is a fixed number, then you would probably say as n approaches infinity, $b^{\left(\frac{1}{n}\right)}$ approaches $b^0 = 1$. After all, we did already prove that $\lim_{n\to\infty} \frac{1}{n} = 0$. We should be able to back up this intuition with our rigorous definition.

Problem 6.1.13 Let $b > 0$. Use the definition to prove $\lim\limits_{n\to\infty} b^{\left(\frac{1}{n}\right)} = 1$.

Hint. You will probably need to separate this into two cases: $0 < b < 1$ and $b \geq 1$. □

Problem 6.1.14

(a) Provide a rigorous definition for $\lim\limits_{n\to\infty} s_n \neq s$.

(b) Use your definition to show that for any real number a, $\lim\limits_{n\to\infty} \left((-1)^n\right) \neq a$.

Hint. Choose $\varepsilon = 1$ and use the fact that $\left|a - (-1)^n\right| < 1$ is equivalent to $(-1)^n - 1 < a < (-1)^n + 1$ to show that no choice of N will work for this ε.

□

6.2 The Limit as a Primary Tool

As you've seen from the previous sections, the formal definition of the convergence of a sequence is meant to capture rigorously our intuitive understanding of convergence. However, the definition itself is an unwieldy tool. If only there was a way to be rigorous without having to run back to the definition each time. Fortunately, there is a way. If we can use the definition to prove some general rules about limits then we could use these rules whenever they applied and be assured that everything was still rigorous. A number of these should look familiar from calculus.

Problem 6.2.1 Let $(c)_{n=1}^{\infty} = (c, c, c, \ldots)$ be a constant sequence. Show that $\lim_{n\to\infty} c = c$. □

In proving the familiar limit theorems, the following will prove to be a very useful tool.

Lemma 6.2.2

Triangle Inequality: *Let a and b be real numbers. Then*

$$|a + b| \leq |a| + |b|.$$

Reverse Triangle Inequality: *Let a and b be real numbers. Then*

$$|a| - |b| \leq |a - b|$$

Problem 6.2.3

(a) Prove Lemma 6.2.2.

Hint.

For the Reverse Triangle Inequality, consider $|a| = |a - b + b|$.

(b) Show $||a| - |b|| \leq |a - b|$.

Hint.

You want to show $|a| - |b| \leq |a - b|$ and $-(|a| - |b|) \leq |a - b|$.

□

Theorem 6.2.4 *If $\lim_{n\to\infty} a_n = a$ and $\lim_{n\to\infty} b_n = b$, then $\lim_{n\to\infty} (a_n + b_n) = a + b$.*

We will often informally state this theorem as "the limit of a sum is the sum of the limits." However, to be absolutely precise, what it says is that if we already know that two sequences converge, then the sequence formed by summing the corresponding terms of those two sequences will converge and, in fact, converge to the sum of those individual limits. We'll provide the scrapwork for the proof of this and leave the formal write-up as an exercise. Note the use of the triangle inequality in the proof.

Problem 6.2.5 Prove Theorem 6.2.4.

SCRAPWORK:

If we let $\varepsilon > 0$, then we want N so that if $n > N$, then $\left|(a_n + b_n) - (a + b)\right| < \varepsilon$. We *know* that $\lim_{n\to\infty} a_n = a$ and $\lim_{n\to\infty} b_n = b$, so we can make $\left|a_n - a\right|$ and $\left|b_n - b\right|$ as small as we wish, provided we make n large enough. Let's go back to what we want, to see if we can close the gap between what we know and what we want.

$$|(a_n + b_n) - (a + b)| = |(a_n - a) + (b_n - b)| \leq |a_n - a| + |b_n - b|$$

by the triangle inequality. To make this whole thing less than ε, it makes sense to make each part less than $\frac{\varepsilon}{2}$. it makes sense to make each part less than $\frac{\varepsilon}{2}$. Fortunately, we can do that as the definitions of $\lim_{n\to\infty} a_n = a$ and $\lim_{n\to\infty} b_n = b$ allow us to make $\left|a_n - a\right|$ and $\left|b_n - b\right|$ arbitrarily small. Specifically, since $\lim_{n\to\infty} a_n = a$, there exists an N_1 such that if $n > N_1$ then $\left|a_n - a\right| < \frac{\varepsilon}{2}$. Also since $\lim_{n\to\infty} b_n = b$, there exists an N_2 such that if $n > N_2$ then $\left|b_n - b\right| < \frac{\varepsilon}{2}$. Since we want both of these to occur, it makes sense to let $N = \max(N_1, N_2)$. This should be the N that we seek. □

Theorem 6.2.6 *If $\lim_{n\to\infty} a_n = a$ and $\lim_{n\to\infty} b_n = b$, then $\lim_{n\to\infty} (a_n b_n) = ab$.*

SCRAPWORK: Given $\varepsilon > 0$, we want N so that if $n > N$, then $\left|a_n b_n - ab\right| < \varepsilon$. One of the standard tricks in analysis is to "uncancel." In this case we will subtract and add a convenient term. Normally these would "cancel out," which is why we say that we will uncancel to put them back in. You already saw an example of this in proving the Reverse Triangle Inequality (Problem 6.2.3). In the present case, consider

$$|a_n b_n - ab| = |a_n b_n - a_n b + a_n b - ab|$$

$$\leq |a_n b_n - a_n b| + |a_n b - ab|$$
$$= |a_n|\,|b_n - b| + |b|\,|a_n - a|\,.$$

We can make this whole thing less than ε, provided we make each term in the sum less than $\frac{\varepsilon}{2}$. We can make $|b|\,|a_n - a| < \frac{\varepsilon}{2}$ if we make $|a_n - a| < \frac{\varepsilon}{2|b|}$. But wait! What if $b = 0$? We could handle this as a separate case or we can do the following "slick trick." Notice that we can add one more line to the above string of inequalities: $|a_n|\,|b_n - b| + |b|\,|a_n - a| < |a_n|\,|b_n - b| + (|b| + 1)\,|a_n - a|$. Now we can make $|a_n - a| < \frac{\varepsilon}{2(|b|+1)}$ and not worry about dividing by zero.

Making $|a_n|\,|b_n - b| < \frac{\varepsilon}{2}$ requires a bit more finesse. At first glance, one would be tempted to try and make $|b_n - b| < \frac{\varepsilon}{2|a_n|}$. Even if we ignore the fact that we could be dividing by zero (which we could handle), we have a bigger problem. According to the definition of $\lim_{n\to\infty} b_n = b$, we can make $|b_n - b|$ smaller than any given fixed positive number, as long as we make n large enough (larger than some N which goes with a given epsilon). Unfortunately, $\frac{\varepsilon}{2|a_n|}$ is not fixed as it has the variable n in it; there is no reason to believe that a single N will work with all of these simultaneously. To handle this impasse, we need the following:

Lemma 6.2.7 ***A Convergent Sequence Is Bounded.***
If $\lim_{n\to\infty} a_n = a$, then there exists $B > 0$ such that $|a_n| \leq B$ for all n.

Problem 6.2.8 Prove Lemma 6.2.7.

Hint.
We know that there exists N such that if $n > N$, then $|a_n - a| < 1$. Let $B = \max\left(|a_1|, |a_2|, \ldots, \left|a_{\lceil N\rceil}\right|, |a| + 1\right)$, where $\lceil N\rceil$ represents the smallest integer greater than or equal to N. Also, notice that this is not a convergence proof so it is not safe to think of N as a large number.

Comment. Actually, this is a dangerous habit to fall into even in convergence proofs.

□

Armed with this bound B, we can add on one more inequality to the above scrapwork to get

$$\begin{aligned}|a_n \cdot b_n - a \cdot b| &= |a_n \cdot b_n - a_n \cdot b + a_n \cdot b - a \cdot b| \\ &\leq |a_n \cdot b_n - a_n \cdot b| + |a_n \cdot b - a \cdot b| \\ &= |a_n|\,|b_n - b| + |b|\,|a_n - a| \\ &< B\,|b_n - b| + (|b| + 1)\,|a_n - a|\end{aligned}$$

At this point, we should be able to make the last line of this less than ε.
END OF SCRAPWORK

Problem 6.2.9 Prove Theorem 6.2.6. □

Corollary 6.2.10 *(Corollary to Theorem 6.2.6) If $\lim\limits_{n\to\infty} a_n = a$ and $c \in \mathbb{R}$, then $\lim\limits_{n\to\infty} c \cdot a_n = c \cdot a$.*

Problem 6.2.11 Prove Corollary 6.2.10. □

Just as Theorem 6.2.6 says that the limit of a product is the product of the limits, we can prove the analogue for quotients.

Theorem 6.2.12 *Suppose $\lim\limits_{n\to\infty} a_n = a$ and $\lim\limits_{n\to\infty} b_n = b$. Also suppose $b \neq 0$ and $b_n \neq 0, \forall\, n$. Then $\lim\limits_{n\to\infty} \left(\dfrac{a_n}{b_n}\right) = \dfrac{a}{b}$.*

Problem 6.2.13 Prove Theorem 6.2.12.

SCRAPWORK

To prove this, let's look at the special case of trying to prove $\lim_{n\to\infty}\left(\frac{1}{b_n}\right) = \frac{1}{b}$. The general case will follow from this and Theorem 6.2.6. Consider $\left|\frac{1}{b_n} - \frac{1}{b}\right| = \frac{|b-b_n|}{|b_n||b|}$. We are faced with the same dilemma as before; we need to get $\left|\frac{1}{b_n}\right|$ bounded above. This means we need to get $|b_n|$ bounded away from zero (at least for large enough n).

This can be done as follows. Since $b \neq 0$, then $\frac{|b|}{2} > 0$. Thus, by the definition of $\lim_{n\to\infty} b_n = b$, there exists N_1 such that if $n > N_1$, then $|b|-|b_n| \leq \left|b - b_n\right| < \frac{|b|}{2}$. Thus when $n > N_1$, $\frac{|b|}{2} < |b_n|$ and so $\frac{1}{|b_n|} < \frac{2}{|b|}$. This says that for $n > N_1$, $\frac{|b-b_n|}{|b_n||b|} < \frac{2}{|b|^2}|b - b_n|$. We should be able to make this smaller than a given $\varepsilon > 0$, provided we make n large enough. □

These theorems allow us to compute limits of complicated sequences and rigorously verify that these are, in fact, the correct limits without resorting to the definition of a limit.

Problem 6.2.14 Identify all of the theorems implicitly used to show that

$$\lim_{n\to\infty} \frac{3n^3 - 100n + 1}{5n^3 + 4n^2 - 7} = \lim_{n\to\infty} \frac{n^3\left(3 - \frac{100}{n^2} + \frac{1}{n^3}\right)}{n^3\left(5 + \frac{4}{n} - \frac{7}{n^3}\right)} = \frac{3}{5}.$$

Notice that this presumes that all of the individual limits exist. This will become evident as the limit is decomposed. □

There is one more tool that will prove to be valuable.

Theorem 6.2.15 ***Squeeze Theorem for Sequences***

Let $(r_n), (s_n)$, and (t_n) be sequences of real numbers with $r_n \leq s_n \leq t_n, \forall$ positive integers n. Suppose $\lim_{n\to\infty} r_n = s = \lim_{n\to\infty} t_n$. Then (s_n) must converge and $\lim_{n\to\infty} s_n = s$.

Problem 6.2.16 Prove Theorem 6.2.15.

Hint. This is probably a place where you would want to use $s-\varepsilon < s_n < s+\varepsilon$ instead of $|s_n - s| < \varepsilon$. □

The Squeeze Theorem holds even if $r_n \leq s_n \leq t_n$ holds for only sufficiently large n; i.e., for n larger than some fixed N_0. This is true because when you find an N_1 that works in the original proof, this can be modified by choosing $N = \max(N_0, N_1)$. Also note that this theorem really says two things: (s_n) converges and it converges to s. This subtle point affects how one should properly use the Squeeze Theorem.

Example 6.2.17 Prove $\lim\limits_{n\to\infty} \dfrac{n+1}{n^2} = 0.$ □

Proof. Notice that $0 \leq \frac{n+1}{n^2} \leq \frac{n+n}{n^2} = \frac{2}{n}$. Since $\lim\limits_{n\to\infty} 0 = 0 = \lim\limits_{n\to\infty} \dfrac{2}{n}$, then by the Squeeze Theorem, $\lim\limits_{n\to\infty} \dfrac{n+1}{n^2} = 0.$ ■

Notice that this proof is completely rigorous. Also notice that this is the proper way to use the Squeeze Theorem. Here is an example of an {improper} use of the Squeeze Theorem.

How *not* to prove Example 6.2.17. Notice that

$$0 \leq \frac{n+1}{n^2} \leq \frac{n+n}{n^2} = \frac{2}{n},$$

so

$$0 = \lim_{n\to\infty} 0 \leq \lim_{n\to\infty} \frac{n+1}{n^2} \leq \lim_{n\to\infty} \frac{2}{n} = 0$$

and

$$\lim_{n\to\infty} \frac{n+1}{n^2} = 0.$$

This is incorrect in form because it presumes that $\lim_{n\to\infty} \frac{n+1}{n^2}$ exists, which we don't yet know. If we knew that the limit existed to begin with, then this would be fine. The Squeeze Theorem proves that the limit does in fact exist, but it must be so stated.

These general theorems will allow us to rigorously explore convergence of power series in the next chapter without having to appeal directly to the definition of convergence. However, you should remember that we used the definition to prove these results and there will be times when we will need to apply the definition directly. However, before we go into that, let's examine divergence a bit more closely.

6.3 Divergence

In Theorem 4.2.2 we saw that there is a rearrangment of the alternating Harmonic series which diverges to ∞ or $-\infty$. In that section we did not fuss over any formal notions of divergence. We assumed instead that you are already familiar with the concept of divergence, probably from taking calculus in the past.

However we are now in the process of building precise, formal definitions for the concepts we will be using so we define the divergence of a sequence as follows.

Definition 6.3.1 A sequence of real numbers $(s_n)_{n=1}^{\infty}$ diverges if it does not converge to any $a \in \mathbb{R}$. ♢

It may seem unnecessarily pedantic of us to insist on formally stating such an obvious definition. After all "converge" and "diverge" are opposites in ordinary English. Why wouldn't they be mathematically opposite too? Why do we have to go to the trouble of formally defining both of them? Since they are opposites defining one implicitly defines the other doesn't it?

One way to answer that criticism is to state that in mathematics we *always* work from precisely stated definitions and tightly reasoned logical arguments.

But this is just more pedantry. It is a way of saying, "Because we said so" all dressed up in imposing language. We need to do better than that.

One reason for providing formal definitions of both convergence and divergence is that in mathematics we frequently co-opt words from natural languages like English and imbue them with mathematical meaning that is only tangentially related to the original English definition. When we take two such words which happen to be opposites in English and give them mathematical meanings which are *not* opposites it can be very confusing, especially at first.

This is what happened with the words "open" and "closed." These are opposites in English: "not open" is "closed," "not closed" is "open," and there is nothing which is both open and closed. But recall that an open interval on the real line, (a, b), is one that does not include *either* of its endpoints while a closed interval, $[a, b]$, is one that includes both of them.

These may seem like opposites at first but they are not. To see this observe that the interval $(a, b]$ is neither open nor closed since it only contains one of its endpoints. If "open" and "closed" were mathematically opposite then *every* interval would be either open or closed.

Open Sets vs. Closed Sets. It is also true that $(-\infty, \infty)$ is *both* open and closed, but an explanation of this would take us too far afield.

Mathematicians have learned to be extremely careful about this sort of thing. In the case of convergence and divergence of a series, even though these words are actually opposites mathematically (every sequence either converges or diverges and no sequence converges *and* diverges) it is better to say this explicitly so there can be no confusion.

A sequence $(a_n)_{n=1}^{\infty}$ can only converge to a real number, a, in one way: by getting arbitrarily close to a. However there are several ways a sequence might diverge.

Example 6.3.2 Consider the sequence, $(n)_{n=1}^{\infty}$. This clearly diverges by getting larger and larger ... Ooops! Let's be careful. The sequence $\left(1-\frac{1}{n}\right)_{n=1}^{\infty}$ gets larger and larger too, but it converges. What we meant to say was that the terms of the sequence $(n)_{n=1}^{\infty}$ become *arbitrarily* large as n increases.

This is clearly a divergent sequence but it may not be clear how to prove this formally. Here's one way.

To show divergence we must show that the sequence satisfies the *negation* of the definition of convergence. That is, we must show that for every $r \in \mathbb{R}$ there is an $\varepsilon > 0$ such that for every $N \in \mathbb{R}$, there is an $n > N$ with $|n - r| \geq \varepsilon$.

So let $\varepsilon = 1$, and let $r \in \mathbb{R}$ be given. Let $N = r + 2$. Then for every $n > N$ $|n - r| > |(r + 2) - r| = 2 > 1$. Therefore the sequence diverges. □

This seems to have been rather more work than we should have to do for such a simple problem. Here's another way which highlights this particular type of divergence.

First we'll need a new definition:

Definition 6.3.3 A sequence, $(a_n)_{n=1}^{\infty}$, **diverges to positive infinity** if for every real number r, there is a real number N such that $n > N \Rightarrow a_n > r$.

A sequence, $(a_n)_{n=1}^{\infty}$, **diverges to negative infinity** if for every real number r, there is a real number N such that $n > N \Rightarrow a_n < r$.

A sequence is said to **diverge to infinity** if it diverges to either positive or negative infinity. ◊

In practice we want to think of $|r|$ as a very large number. This definition says that a sequence diverges to infinity if it becomes arbitrarily large as n increases, and similarly for divergence to negative infinity.

Problem 6.3.4 Show that $(n)_{n=1}^{\infty}$ diverges to infinity. □

Problem 6.3.5 Show that if $(a_n)_{n=1}^{\infty}$ diverges to infinity then $(a_n)_{n=1}^{\infty}$ diverges. □

We will denote divergence to infinity as

$$\lim_{n\to\infty} a_n = \pm\infty.$$

However, strictly speaking this is an abuse of notation since the symbol ∞ does not represent a real number. This notation can be very problematic since it looks so much like the notation we use to denote convergence: $\lim_{n\to\infty} a_n = a$.

Nevertheless, the notation is appropriate because divergence to infinity is "nice" divergence in the sense that it shares many of the properties of convergence, as the next problem shows.

Problem 6.3.6 Suppose $\lim_{n\to\infty} a_n = \infty$ and $\lim_{n\to\infty} b_n = \infty$.

(a) Show that $\lim_{n\to\infty} a_n + b_n = \infty$

(b) Show that $\lim_{n\to\infty} a_n b_n = \infty$

(c) Is it true that $\lim_{n\to\infty} \frac{a_n}{b_n} = \infty$? Explain.

□

Because divergence to positive or negative infinity shares some of the properties of convergence it is easy to get careless with it. Remember that even though we write $\lim_{n\to\infty} a_n = \infty$ this is still a divergent sequence in the sense that $\lim_{n\to\infty} a_n$ does not exist. The symbol ∞ does not represent a real number. This is just a convenient notational shorthand telling us that the sequence diverges by becoming arbitrarily large.

Problem 6.3.7 Suppose $\lim_{n\to\infty} a_n = \infty$ and $\lim_{n\to\infty} b_n = -\infty$ and $\alpha \in \mathbb{R}$. Prove or give a counterexample:

(a) $\lim_{n\to\infty} a_n + b_n = \infty$

(b) $\lim_{n\to\infty} a_n b_n = -\infty$

(c) $\lim_{n\to\infty} \alpha a_n = \infty$

(d) $\lim_{n\to\infty} \alpha b_n = -\infty$

□

Finally, a sequence can diverge in other ways as the following problem displays.

Problem 6.3.8 Show that each of the following sequences diverge.

(a) $\left((-1)^n\right)_{n=1}^{\infty}$

(b) $\left((-1)^n\, n\right)_{n=1}^{\infty}$

(c) $a_n = \begin{cases} 1 & \text{if } n = 2^p \text{ for some } p \in \mathbb{N} \\ \frac{1}{n} & \text{otherwise.} \end{cases}$

□

Problem 6.3.9 Suppose that $(a_n)_{n=1}^{\infty}$ diverges but not to infinity and that α is a real number. What conditions on α will guarantee that:

(a) $(\alpha a_n)_{n=1}^{\infty}$ converges?

(b) $(\alpha a_n)_{n=1}^{\infty}$ diverges?

□

Problem 6.3.10 Show that if $|r| > 1$ then $(r^n)_{n=1}^{\infty}$ diverges. Will it diverge to infinity? □

6.4 Additional Problems

Problem 6.4.1 Prove that if $\lim_{n\to\infty} s_n = s$ then $\lim_{n\to\infty} |s_n| = |s|$. Prove that the converse is true when $s = 0$, but it is not necessarily true otherwise.

□

Problem 6.4.2

(a) Let (s_n) and (t_n) be sequences with $s_n \le t_n, \forall n$. Suppose $\lim_{n\to\infty} s_n = s$ and $\lim_{n\to\infty} t_n = t$.

Prove $s \leq t$.

Hint.

Assume for contradiction, that $s > t$ and use the definition of convergence with $\varepsilon = \frac{s-t}{2}$ to produce an n with $s_n > t_n$.

(b) Prove that if a sequence converges, then its limit is unique. That is, prove that if $\lim_{n\to\infty} s_n = s$ and $\lim_{n\to\infty} s_n = t$, then $s = t$.

□

Problem 6.4.3 Prove that if the sequence (s_n) is bounded then $\lim_{n\to\infty}\left(\frac{s_n}{n}\right) = 0$. □

Problem 6.4.4

(a) Prove that if $x \neq 1$, then

$$1 + x + x^2 + \cdots + x^n = \frac{1 - x^{n+1}}{1 - x}.$$

(b) Use (a) to prove that if $|x| < 1$, then $\lim_{n\to\infty}\left(\sum_{j=0}^{n} x^j\right) = \frac{1}{1-x}$.

□

Problem 6.4.5 Prove

$$\lim_{n\to\infty} \frac{a_0 + a_1 n + a_2 n^2 + \cdots + a_k n^k}{b_0 + b_1 n + b_2 n^2 + \cdots + b_k n^k} = \frac{a_k}{b_k},$$

provided $b_k \neq 0$. [Notice that since a polynomial only has finitely many roots, then the denominator will be non-zero when n is sufficiently large.] □

Problem 6.4.6 Prove that if $\lim_{n\to\infty} s_n = s$ and $\lim_{n\to\infty}(s_n - t_n) = 0$, then $\lim_{n\to\infty} t_n = s$. □

Problem 6.4.7

(a) Prove that if $\lim_{n\to\infty} s_n = s$ and $s < t$, then there exists a real number N such that if $n > N$ then $s_n < t$.

(b) Prove that if $\lim_{n\to\infty} s_n = s$ and $r < s$, then there exists a real number M such that if $n > M$ then $r < s_n$.

□

Problem 6.4.8 Suppose (s_n) is a sequence of positive numbers such that

$$\lim_{n\to\infty}\left(\frac{s_{n+1}}{s_n}\right) = L.$$

(a) Prove that if $L < 1$, then $\lim_{n\to\infty} s_n = 0$.

Hint.

Choose R with $L < R < 1$. By the previous problem, $\exists\ N$ such that if $n > N$, then $\frac{s_{n+1}}{s_n} < R$. Let $n_0 > N$ be fixed and show $s_{n_0+k} < R^k s_{n_0}$. Conclude that $\lim_{k\to\infty} s_{n_0+k} = 0$ and let $n = n_0 + k$.

(b) Let c be a positive real number. Prove $\lim\limits_{n\to\infty}\left(\frac{c^n}{n!}\right) = 0$.

□

Chapter 7

A "Tayl" of Three Remainders

7.1 The Integral Form of the Remainder

Now that we have a rigorous definition of the convergence of a sequence, let's apply this to Taylor series. Recall that the Taylor series of a function $f(x)$ expanded about the point a is given by

$$\sum_{n=0}^{\infty} \frac{f^{(n)}(a)}{n!}(x-a)^n = f(a) + \frac{f'(a)}{1!}(x-a) + \frac{f''(a)}{2!}(x-a)^2 + \cdots$$

When we say that $f(x) = \sum_{n=0}^{\infty} \frac{f^{(n)}(a)}{n!}(x-a)^n$ for a particular value of x, what we mean is that the sequence of partial sums

$$\left(\sum_{j=0}^{n} \frac{f^{(j)}(a)}{j!}(x-a)^j\right)_{n=0}^{\infty}$$
$$= \left(f(a), f(a) + \frac{f'(a)}{1!}(x-a), f(a) + \frac{f'(a)}{1!}(x-a) + \frac{f''(a)}{2!}(x-a)^2, \ldots\right)$$

converges to the number $f(x)$. Note that the index in the summation was changed to j to allow n to represent the index of the sequence of partial sums. As intimidating as this may look, bear in mind that for a fixed real number x, this is still a sequence of real numbers so, that saying $f(x) = \sum_{n=0}^{\infty} \frac{f^{(n)}(a)}{n!}(x-a)^n$ means that $\lim_{n\to\infty}\left(\sum_{j=0}^{n} \frac{f^{(j)}(a)}{j!}(x-a)^j\right) = f(x)$ and in the previous chapter we developed some tools to examine this phenomenon. In particular, we know that $\lim_{n\to\infty}\left(\sum_{j=0}^{n} \frac{f^{(j)}(a)}{j!}(x-a)^j\right) = f(x)$ is equivalent to

$$\lim_{n\to\infty}\left[f(x) - \left(\sum_{j=0}^{n} \frac{f^{(j)}(a)}{j!}(x-a)^j\right)\right] = 0.$$

We have seen an example of this already. Problem 6.4.4 of the last chapter basically had you show that the geometric series, $1 + x + x^2 + x^3 + \cdots$ converges to $\frac{1}{1-x}$,for $|x| < 1$ by showing that $\lim_{n\to\infty}\left[\frac{1}{1-x} - \left(\sum_{j=0}^{n} x^j\right)\right] = 0$.

There is generally not a readily recognizable closed form for the partial sum for a Taylor series. The geometric series is a special case. Fortunately, for the issue at hand (convergence of a Taylor series), we don't need to analyze the series itself. What we need to show is that the difference between the function and the nth partial sum converges to zero. This difference is called the **remainder (of the Taylor series)**. (Why?)

While it is true that the remainder is simply

$$f(x) - \left(\sum_{j=0}^{n} \frac{f^{(j)}(a)}{j!}(x-a)^j\right),$$

this form is not easy to work with. Fortunately, a number of alternate versions of this remainder are available. We will explore these in this chapter.

Recall the result from Theorem 4.1.9 from Chapter 4,

$$f(x) = f(a) + \frac{f'(a)}{1!}(x-a) + \frac{f''(a)}{2!}(x-a)^2 + \cdots + \frac{f^{(n)}(a)}{n!}(x-a)^n + \frac{1}{n!}\int_{t=a}^{x} f^{(n+1)}(t)(x-t)^n\, dt.$$

We can use this by rewriting it as

$$f(x) - \left(\sum_{j=0}^{n} \frac{f^{(j)}(a)}{j!}(x-a)^j\right) = \frac{1}{n!}\int_{t=a}^{x} f^{(n+1)}(t)(x-t)^n\, dt.$$

The expression $\frac{1}{n!}\int_{t=a}^{x} f^{(n+1)}(t)(x-t)^n\, dt$ is called the *integral form of the remainder for the Taylor series* of $f(x)$, and the Taylor series will converge to $f(x)$ exactly when the sequence $\lim_{n\to\infty}\left(\frac{1}{n!}\int_{t=a}^{x} f^{(n+1)}(t)(x-t)^n\, dt\right)$ converges to zero. It turns out that this form of the remainder is often easier to handle than the original $f(x) - \left(\sum_{j=0}^{n} \frac{f^{(j)}(a)}{j!}(x-a)^j\right)$ and we can use it to obtain some general results.

Theorem 7.1.1 ***Taylor's series***

If there exists a real number B such that $|f^{(n+1)}(t)| \le B$ for all nonnegative integers n and for all t on an interval containing a and x, then

$$\lim_{n\to\infty}\left(\frac{1}{n!}\int_{t=a}^{x} f^{(n+1)}(t)(x-t)^n\, dt\right) = 0$$

and so

$$f(x) = \sum_{n=0}^{\infty} \frac{f^{(n)}(a)}{n!}(x-a)^n.$$

In order to prove this, it might help to first prove the following.

Lemma 7.1.2 *If f and $|f|$ are integrable functions and $a \le b$, then*

$$\left|\int_{t=a}^{b} f(t)\, dt\right| \le \int_{t=a}^{b} |f(t)|\, dt.$$

Problem 7.1.3 Prove Lemma 7.1.2.

Hint.

$-|f(t)| \le f(t) \le |f(t)|$. □

Problem 7.1.4 Prove Theorem 7.1.1.

Hint.

You might want to use Problem 6.4.8 of Chapter 6. Also there are two cases to consider: $a < x$ and $x < a$ (the case $x = a$ is trivial). You will find that this is true in general. This is why we will often indicate that t is between a and x as in the theorem. In the case $x < a$, notice that

$$\left|\int_{t=a}^{x} f^{(n+1)}(t)(x-t)^n \, \mathrm{d}t\right| = \left|(-1)^{n+1}\int_{t=x}^{a} f^{(n+1)}(t)(t-x)^n \, \mathrm{d}t\right| = \left|\int_{t=x}^{a} f^{(n+1)}(t)(t-x)^n \, \mathrm{d}t\right|.$$

□

Problem 7.1.5 Use Theorem 7.1.1 to prove that for any real number x

(a) $\sin x = \sum_{n=0}^{\infty} \frac{(-1)^n x^{2n+1}}{(2n+1)!}$

(b) $\cos x = \sum_{n=0}^{\infty} \frac{(-1)^n x^{2n}}{(2n)!}$

(c) $e^x = \sum_{n=0}^{\infty} \frac{x^n}{n!}$

□

Problem 7.1.5.c shows that the Taylor series of e^x expanded at zero converges to e^x for any real number x. Theorem 7.1.1 can be used in a similar fashion to show that

$$e^x = \sum_{n=0}^{\infty} \frac{e^a (x-a)^n}{n!}$$

for any real numbers a and x.

Recall that Chapter 3 we showed that if we define the function $E(x)$ by the power series $\sum_{n=0}^{\infty} \frac{x^n}{n!}$ then $E(x+y) = E(x)E(y)$. This, of course, is just the familiar addition property of integer exponents extended to any real number. In Chapter 3 we had to *assume* that defining $E(x)$ as a series was meaningful because we did not address the convergence of the series in that chapter. Now that we know the series converges for any real number we see that the definition

$$f(x) = e^x = \sum_{n=0}^{\infty} \frac{x^n}{n!}$$

is in fact valid.

Assuming that we can differentiate this series term-by-term it is straightforward to show that $f'(x) = f(x)$.

Comment. We can, but that proof will have to wait for a few more chapters.

Along with Taylor's formula this can then be used to show that $e^{a+b} = e^a e^b$ more elegantly than the rather cumbersome proof in equation (I.6), as the following problem shows.

Problem 7.1.6 Recall that if $f(x) = e^x$ then $f'(x) = e^x$. Use this along with the Taylor series expansion of e^x about a to show that

$$e^{a+b} = e^a e^b.$$

□

Theorem 7.1.1 is a nice "first step" toward a rigorous theory of the convergence of Taylor series, but it is not applicable in all cases. For example, consider the function $f(x) = \sqrt{1+x}$. As we saw in Chapter 3, Problem 3.2.12, this function's Maclaurin series (the binomial series for $(1+x)^{1/2}$)appears to be converging to the function for $x \in (-1, 1)$. While this is, in fact, true, the above proposition does not apply. If we consider the derivatives of $f(t) = (1+t)^{1/2}$, we obtain:

$$
\begin{aligned}
f'(t) &= \frac{1}{2}(1+t)^{\frac{1}{2}-1} \\
f''(t) &= \frac{1}{2}\left(\frac{1}{2}-1\right)(1+t)^{\frac{1}{2}-2} \\
f'''(t) &= \frac{1}{2}\left(\frac{1}{2}-1\right)\left(\frac{1}{2}-2\right)(1+t)^{\frac{1}{2}-3} \\
&\vdots \\
f^{(n+1)}(t) &= \frac{1}{2}\left(\frac{1}{2}-1\right)\left(\frac{1}{2}-2\right)\cdots\left(\frac{1}{2}-n\right)(1+t)^{\frac{1}{2}-(n+1)}.
\end{aligned}
$$

Notice that

$$\left|f^{(n+1)}(0)\right| = \frac{1}{2}\left(1-\frac{1}{2}\right)\left(2-\frac{1}{2}\right)\cdots\left(n-\frac{1}{2}\right).$$

Since this sequence grows without bound as $n \to \infty$, then there is no chance for us to find a number B to act as a bound for all of the derviatives of f on any interval containing 0 and x, and so the hypothesis of Theorem 7.1.1 will never be satisfied. We need a more delicate argument to prove that

$$\sqrt{1+x} = 1 + \frac{1}{2}x + \frac{\frac{1}{2}\left(\frac{1}{2}-1\right)}{2!}x^2 + \frac{\frac{1}{2}\left(\frac{1}{2}-1\right)\left(\frac{1}{2}-2\right)}{3!}x^3 + \cdots$$

is valid for $x \in (-1, 1)$. To accomplish this task, we will need to express the remainder of the Taylor series differently. Fortunately, there are at least two such alternate forms.

7.2 Lagrange's Form of the Remainder

Joseph-Louis Lagrange provided an alternate form for the remainder in Taylor series in his 1797 work *Théorie des functions analytiques*. Lagrange's form of the remainder is as follows.

Theorem 7.2.1 ***Lagrange's Form of the Remainder***

Suppose f is a function such that $f^{(n+1)}(t)$ is continuous on an interval containing a and x. Then

$$f(x) - \left(\sum_{j=0}^{n} \frac{f^{(j)}(a)}{j!}(x-a)^j\right) = \frac{f^{(n+1)}(c)}{(n+1)!}(x-a)^{n+1}$$

where c is some number between a and x.

Proof. Note first that the result is true when $x = a$ as both sides reduce to 0 (in that case $c = x = a$.) We will prove the case where $a < x$; the case $x < a$ will be an exercise.

First, we already have

$$f(x)-\left(\sum_{j=0}^{n}\frac{f^{(j)}(a)}{j!}(x-a)^j\right)=\frac{1}{n!}\int_{t=a}^{x}f^{(n+1)}(t)(x-t)^n\,\mathrm{d}t$$

so it suffices to show that

$$\int_{t=a}^{x}f^{(n+1)}(t)(x-t)^n\,\mathrm{d}t=\frac{f^{(n+1)}(c)}{n+1}(x-a)^{n+1}$$

for some c with $c\in[a,x]$. To this end, let

$$M=\max_{a\le t\le x}\left(f^{(n+1)}(t)\right)$$

and

$$m=\min_{a\le t\le x}\left(f^{(n+1)}(t)\right).$$

Note that for all $t\in[a,x]$, we have $m\le f^{(n+1)}(t)\le M$. Since $x-t\ge 0$, this gives us

$$m\,(x-t)^n\le f^{(n+1)}(t)(x-t)^n\le M(x-t)^n \tag{III.1}$$

and so

$$\int_{t=a}^{x}m\,(x-t)^n\,\mathrm{d}t\le\int_{t=a}^{x}f^{(n+1)}(t)(x-t)^n\,\mathrm{d}t\le\int_{t=a}^{x}M(x-t)^n\,\mathrm{d}t. \tag{III.2}$$

Computing the outside integrals, we have

$$m\int_{t=a}^{x}(x-t)^n\,\mathrm{d}t\le\int_{t=a}^{x}f^{(n+1)}(t)(x-t)^n\,\mathrm{d}t\le M\int_{t=a}^{x}(x-t)^n\,\mathrm{d}t$$

$$m\frac{(x-a)^{n+1}}{n+1}\le\int_{t=a}^{x}f^{(n+1)}(t)(x-t)^n\,\mathrm{d}t\le M\frac{(x-a)^{n+1}}{n+1} \tag{III.3}$$

$$m\le\frac{\int_{t=a}^{x}f^{(n+1)}(t)(x-t)^n\,\mathrm{d}t}{\left(\frac{(x-a)^{n+1}}{n+1}\right)}\le M.$$

Since

$$\frac{\int_{t=a}^{x}f^{(n+1)}(t)(x-t)^n\,\mathrm{d}t}{\left(\frac{(x-a)^{n+1}}{n+1}\right)}$$

is a value that lies between the maximum and minimum of $f^{(n+1)}$ on $[a,x]$, then by the Intermediate Value Theorem, there must exist a number $c\in[a,x]$ with

$$f^{(n+1)}(c)=\frac{\int_{t=a}^{x}f^{(n+1)}(t)(x-t)^n\,\mathrm{d}t}{\left(\frac{(x-a)^{n+1}}{n+1}\right)}.$$

This gives us

$$\int_{t=a}^{x}f^{(n+1)}(t)(x-t)^n\,\mathrm{d}t=\frac{f^{(n+1)}(c)}{n+1}(x-a)^{n+1}.$$

And the result follows. ■

Problem 7.2.2 Prove Theorem 7.2.1 for the case where $x<a$.

Hint.

Note that

$$\int_{t=a}^{x} f^{(n+1)}(t)(x-t)^n \, dt = (-1)^{n+1} \int_{t=x}^{a} f^{(n+1)}(t)(t-x)^n \, dt.$$

Use the same argument on this integral. It will work out in the end. Really! You just need to keep track of *all* of the negatives. □

This is *not* Lagrange's proof. He did not use the integral form of the remainder. However, this is similar to Lagrange's proof in that he also used the Intermediate Value Theorem (IVT) and Extreme Value Theorem (EVT) much as we did. In Lagrange's day, these were taken to be obviously true for a continuous function and we have followed Lagrange's lead by assuming the IVT and the EVT. However, in mathematics we need to keep our assumptions few and simple. The IVT and the EVT do not satisfy this need in the sense that both can be proved from simpler ideas. We will return to this in Chapter 9.

Also, a word of caution about this: Lagrange's form of the remainder is $\frac{f^{(n+1)}(c)}{(n+1)!}(x-a)^{n+1}$, where c is some number between a and x. The proof does not indicate what this c might be and, in fact, this c changes as n changes. All we know is that this c lies between a and x. To illustrate this issue and its potential dangers, consider the following problem where we have a chance to compute the value of c for the function $f(x) = \frac{1}{1+x}$.

Problem 7.2.3 This problem investigates the Taylor series representation

$$\frac{1}{1+x} = 1 - x + x^2 - x^3 + \cdots.$$

(a) Use the fact that $\frac{1-(-x)^{n+1}}{1+x} = 1-x+x^2-x^3+\cdots+(-x)^n$ to compute the remainder $\frac{1}{1+x} - \left(1 - x + x^2 - x^3 + \cdots + (-x)^n\right)$. Specifically, compute this remainder when $x = 1$ and conclude that the Taylor series does not converge to $\frac{1}{1+x}$ when $x = 1$.

(b) Compare the remainder in part a with the Lagrange form of the remainder to determine what c is when $x = 1$.

(c) Consider the following argument: If $f(x) = \frac{1}{1+x}$, then

$$f^{(n+1)}(c) = \frac{(-1)^{n+1}(n+1)!}{(1+c)^{n+2}}$$

so the Lagrange form of the remainder when $x = 1$ is given by

$$\frac{(-1)^{n+1}(n+1)!}{(n+1)!(1+c)^{n+2}} = \frac{(-1)^{n+1}}{(1+c)^{n+2}}$$

where $c \in [0, 1]$. It can be seen in part b that $c \neq 0$. Thus $1 + c > 1$ and so by Problem 6.1.7 of Chapter 6, the Lagrange remainder converges to 0 as $n \to \infty$. This argument would suggest that the Taylor series converges to $\frac{1}{1+x}$ for $x = 1$. However, we know from part (a) that this is incorrect. What is wrong with the argument?

□

Even though there are potential dangers in misusing the Lagrange form of the remainder, it is a useful form. For example, armed with the Lagrange form of the remainder, we can prove the following theorem.

Theorem 7.2.4 *The binomial series*

$$1+\frac{1}{2}x+\frac{\frac{1}{2}\left(\frac{1}{2}-1\right)}{2!}x^2+\frac{\frac{1}{2}\left(\frac{1}{2}-1\right)\left(\frac{1}{2}-2\right)}{3!}x^3+\cdots$$

converges to $\sqrt{1+x}$ *for* $x\in[0,1]$.

Proof. First note that the binomial series is, in fact, the Taylor series for the function $f(x)=\sqrt{1+x}$ expanded about $a=0$. If we let x be a fixed number with $0\le x\le 1$, then it suffices to show that the Lagrange form of the remainder converges to 0. With this in mind, notice that

$$f^{(n+1)}(t)=\left(\frac{1}{2}\right)\left(\frac{1}{2}-1\right)\cdots\left(\frac{1}{2}-n\right)(1+t)^{\frac{1}{2}-(n+1)}$$

and so the Lagrange form of the remainder is

$$\frac{f^{(n+1)}(c)}{(n+1)!}x^{n+1}=\frac{\left(\frac{1}{2}\right)\left(\frac{1}{2}-1\right)\cdots\left(\frac{1}{2}-n\right)}{(n+1)!}\frac{x^{n+1}}{(1+c)^{n+\frac{1}{2}}}$$

where c is some number between 0 and x. Since $0\le x\le 1$ and $1+c\ge 1$, then we have $\frac{1}{1+c}\le 1$, and so

$$\begin{aligned}
0&\le\left|\frac{\left(\frac{1}{2}\right)\left(\frac{1}{2}-1\right)\cdots\left(\frac{1}{2}-n\right)}{(n+1)!}\frac{x^{n+1}}{(1+c)^{n+\frac{1}{2}}}\right|\\
&=\frac{\left(\frac{1}{2}\right)\left(1-\frac{1}{2}\right)\cdots\left(n-\frac{1}{2}\right)}{(n+1)!}\frac{x^{n+1}}{(1+c)^{n+\frac{1}{2}}}\\
&=\frac{\left(\frac{1}{2}\right)\left(\frac{1}{2}\right)\left(\frac{3}{2}\right)\left(\frac{5}{2}\right)\cdots\left(\frac{2n-1}{2}\right)}{(n+1)!}\left(x^{n+1}\right)\frac{1}{(1+c)^{n+\frac{1}{2}}}\\
&\le\frac{1\cdot 1\cdot 3\cdot 5\cdot\cdots\cdot(2n-1)}{2^{n+1}(n+1)!}\\
&=\frac{1\cdot 3\cdot 5\cdot\cdots\cdot(2n-1)\cdot 1}{2\cdot 4\cdot 6\cdot\cdots\cdot 2n\cdot(2n+2)}\\
&=\frac{1}{2}\cdot\frac{3}{4}\cdot\frac{5}{6}\cdot\cdots\cdot\frac{2n-1}{2n}\cdot\frac{1}{2n+2}\\
&\le\frac{1}{2n+2}.
\end{aligned}$$

Since $\lim_{n\to\infty}\frac{1}{2n+2}=0=\lim_{n\to\infty}0$, then by the Squeeze Theorem,

$$\lim_{n\to\infty}\left|\frac{f^{(n+1)}(c)}{(n+1)!}x^{n+1}\right|=0,\text{ so }\lim_{n\to\infty}\left(\frac{f^{(n+1)}(c)}{(n+1)!}x^{n+1}\right)=0.$$

Thus the Taylor series

$$1+\frac{1}{2}x+\frac{\frac{1}{2}\left(\frac{1}{2}-1\right)}{2!}x^2+\frac{\frac{1}{2}\left(\frac{1}{2}-1\right)\left(\frac{1}{2}-2\right)}{3!}x^3+\cdots$$

converges to $\sqrt{1+x}$ for $0\le x\le 1$. ■

Unfortunately, this proof will not work for $-1<x<0$. In this case, the fact that $x\le c\le 0$ makes $1+c\le 1$. Thus $\frac{1}{1+c}\ge 1$ and so the inequality

$$\frac{\left(\frac{1}{2}\right)\left(\frac{1}{2}\right)\left(\frac{3}{2}\right)\left(\frac{5}{2}\right)\cdots\left(\frac{2n-1}{2}\right)}{(n+1)!}\frac{|x|^{n+1}}{(1+c)^{n+\frac{1}{2}}}\le\frac{1\cdot 1\cdot 3\cdot 5\cdot\cdots\cdot(2n-1)}{2^{n+1}(n+1)!}$$

may not hold.

Problem 7.2.5 Show that if $-\frac{1}{2} \leq x \leq c \leq 0$, then $|\frac{x}{1+c}| \leq 1$ and modify the above proof to show that the binomial series converges to $\sqrt{1+x}$ for $x \in [-\frac{1}{2}, 0]$. □

To take care of the case where $-1 < x < -\frac{1}{2}$, we will use yet another form of the remainder for Taylor series. However before we tackle that, we will use the Lagrange form of the remainder to address something mentioned in Chapter 4. Recall that we noticed that the series representation

$$\frac{1}{1+x} = 1 - x + x^2 - x^3 + \cdots$$

did not work when $x = 1$, however we noticed that the series obtained by integrating term by term did seem to converge to the antiderivative of $\frac{1}{1+x}$. Specifically, we have the Taylor series

$$\ln(1+x) = x - \frac{1}{2}x^2 + \frac{1}{3}x^3 - \cdots.$$

Substituting $x = 1$ into this provided the convergent series $1 - \frac{1}{2} + \frac{1}{3} - \frac{1}{4} + \cdots$. We made the claim that this, in fact, converges to $\ln 2$, but that this was not obvious. The Lagrange form of the remainder gives us the machinery to prove this.

Problem 7.2.6

(a) Compute the Lagrange form of the remainder for the Maclaurin series for $\ln(1+x)$.

(b) Show that when $x = 1$, the Lagrange form of the remainder converges to 0 and so the equation $\ln 2 = 1 - \frac{1}{2} + \frac{1}{3} - \frac{1}{4} + \cdots$ is actually correct.

□

7.3 Cauchy's Form of the Remainder

In his 1823 work, *Résumée des leçons données á l'ecole royale polytechnique sur le calcul infintésimal,* Augustin Cauchy provided another form of the remainder for Taylor series.

Theorem 7.3.1 ***Cauchy's Form of the Remainder***
Suppose f is a function such that $f^{(n+1)}(t)$ is continuous on an interval containing a and x. Then

$$f(x) - \left(\sum_{j=0}^{n} \frac{f^{(j)}(a)}{j!}(x-a)^j\right) = \frac{f^{(n+1)}(c)}{n!}(x-c)^n(x-a)$$

where c is some number between a and x.

Problem 7.3.2 Prove Theorem 7.3.1 using an argument similar to the one used in the proof of Theorem 7.2.1. Don't forget there are two cases to consider. □

Using Cauchy's form of the remainder, we can prove that the binomial series

$$1 + \frac{1}{2}x + \frac{\frac{1}{2}\left(\frac{1}{2}-1\right)}{2!}x^2 + \frac{\frac{1}{2}\left(\frac{1}{2}-1\right)\left(\frac{1}{2}-2\right)}{3!}x^3 + \cdots$$

converges to $\sqrt{1+x}$ for $x \in (-1, 0)$.

Comment. Strictly speaking we only need to show this for $x \in (-1, -1/2)$.In problem 7.2.5 we covered$x \in (-1/2, 0)$.

With this in mind, let x be a fixed number with $-1 < x < 0$ and consider that the binomial series is the Maclaurin series for the function $f(x) = (1+x)^{\frac{1}{2}}$. As we saw before,

$$f^{(n+1)}(t) = \left(\frac{1}{2}\right)\left(\frac{1}{2} - 1\right)\cdots\left(\frac{1}{2} - n\right)(1+t)^{\frac{1}{2}-(n+1)},$$

so the Cauchy form of the remainder is given by

$$0 \leq \left|\frac{f^{(n+1)}(c)}{n!}(x-c)^n(x-0)\right| = \left|\frac{\left(\frac{1}{2}\right)\left(\frac{1}{2}-1\right)\cdots\left(\frac{1}{2}-n\right)}{n!}\frac{(x-c)^n}{(1+c)^{n+\frac{1}{2}}}\cdot x\right|$$

where c is some number with $x \leq c \leq 0$. Thus we have

$$\begin{aligned}
0 &\leq \left|\frac{\left(\frac{1}{2}\right)\left(\frac{1}{2}-1\right)\cdots\left(\frac{1}{2}-n\right)}{n!}\frac{(x-c)^n x}{(1+c)^{n+\frac{1}{2}}}\right| \\
&= \frac{\left(\frac{1}{2}\right)\left(1-\frac{1}{2}\right)\cdots\left(n-\frac{1}{2}\right)}{n!}\frac{|x-c|^n|x|}{(1+c)^{n+\frac{1}{2}}} \\
&= \frac{\left(\frac{1}{2}\right)\left(\frac{1}{2}\right)\left(\frac{3}{2}\right)\left(\frac{5}{2}\right)\cdots\left(\frac{2n-1}{2}\right)}{n!}\frac{(c-x)^n}{(1+c)^n}\frac{|x|}{\sqrt{1+c}} \\
&= \frac{1\cdot 1\cdot 3\cdot 5\cdot\cdots\cdot(2n-1)}{2^{n+1}n!}\left(\frac{c-x}{1+c}\right)^n\frac{|x|}{\sqrt{1+c}} \\
&= \frac{1\cdot 1\cdot 3\cdot 5\cdot\cdots\cdot(2n-1)}{2\cdot 2\cdot 4\cdot 6\cdot\cdots\cdot 2n}\left(\frac{c-x}{1+c}\right)^n\frac{|x|}{\sqrt{1+c}} \\
&= \frac{1}{2}\cdot\frac{1}{2}\cdot\frac{3}{4}\cdot\frac{5}{6}\cdot\cdots\cdot\frac{2n-1}{2n}\cdot\left(\frac{c-x}{1+c}\right)^n\frac{|x|}{\sqrt{1+c}} \\
&\leq \left(\frac{c-x}{1+c}\right)^n\frac{|x|}{\sqrt{1+c}}.
\end{aligned}$$

Notice that if $-1 < x \leq c$, then $0 < 1 + x \leq 1 + c$. Thus $0 < \frac{1}{1+c} \leq \frac{1}{1+x}$ and $\frac{1}{\sqrt{1+c}} \leq \frac{1}{\sqrt{1+x}}$. Thus we have

$$0 \leq \left|\frac{\left(\frac{1}{2}\right)\left(\frac{1}{2}-1\right)\cdots\left(\frac{1}{2}-n\right)}{n!}\frac{(x-c)^n x}{(1+c)^{n+\frac{1}{2}}}\right| \leq \left(\frac{c-x}{1+c}\right)^n\frac{|x|}{\sqrt{1+x}}.$$

Problem 7.3.3 Suppose $-1 < x \leq c \leq 0$ and consider the function $g(c) = \frac{c-x}{1+c}$. Show that on $[x, 0]$, g is increasing and use this to conclude that for $-1 < x \leq c \leq 0$,

$$\frac{c-x}{1+c} \leq |x|.$$

Use this fact to finish the proof that the binomial series converges to $\sqrt{1+x}$ for $-1 < x < 0$. □

The proofs of both the Lagrange form and the Cauchy form of the remainder for Taylor series made use of two crucial facts about continuous functions. First, we assumed the Extreme Value Theorem: Any continuous function on a closed

Figure 7.3.4 Augustin Cauchy[14]

bounded interval assumes its maximum and minimum somewhere on the interval. Second, we assumed that any continuous function satisfied the Intermediate Value Theorem: If a continuous function takes on two different values, then it must take on any value between those two values.

Mathematicians in the late 1700s and early 1800s typically considered these facts to be intuitively obvious. This was natural since our understanding of continuity at that time was, solely, intuitive. Intuition is a useful tool, but as we have seen before it is also unreliable. For example consider the following function.

$$f(x) = \begin{cases} x \sin\left(\frac{1}{x}\right), & \text{if } x \neq 0, \\ 0, & \text{if } x = 0 \end{cases}. \tag{III.4}$$

Is this function continuous at 0? Near zero its graph looks like this:

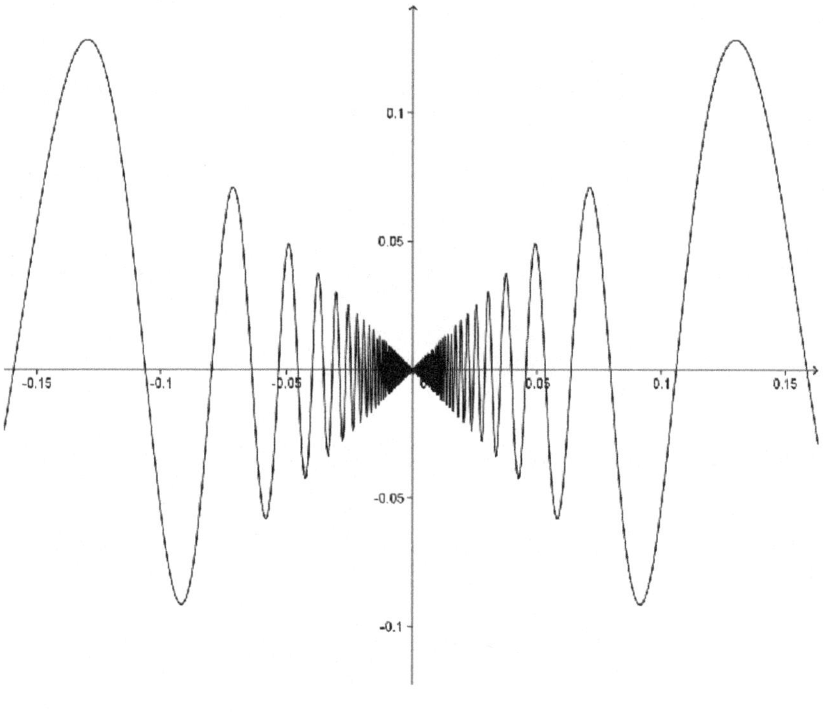

Figure 7.3.5 Formula (III.4) near 0.

but this graph must be taken with a grain of salt as $\sin\left(\frac{1}{x}\right)$ oscillates infinitely often as x nears zero.

[14]mathshistory.st-andrews.ac.uk/Biographies/Cauchy/

No matter what your guess may be, it is clear that it is hard to analyze such a function armed with only an intuitive notion of continuity. We will revisit this example in the next chapter.

As with convergence, continuity is more subtle than it first appears.

We put convergence on solid ground by providing a completely analytic definition in the previous chapter. What we need to do in the next chapter is provide a completely rigorous definition for continuity.

7.4 Additional Problems

Problem 7.4.1 Find the Integral form, Lagrange form, and Cauchy form of the remainder for Taylor series for the following functions expanded about the given values of a.

(a) $f(x) = e^x$, $a = 0$

(b) $f(x) = \sqrt{x}$, $a = 1$

(c) $f(x) = (1 + x)^\alpha$, $a = 0$

(d) $f(x) = \frac{1}{x}$, $a = 3$

(e) $f(x) = \ln x$, $a = 2$

(f) $f(x) = \cos x, a = \frac{\pi}{2}$

□

Chapter 8

Continuity: What It Isn't and What It Is

8.1 An Analytic Definition of Continuity

Before the invention of calculus, the notion of continuity was treated intuitively if it was treated at all. At first pass, it seems a very simple idea based solidly in our experience of the real world. Standing on the bank we see a river flow past us continuously, not by tiny jerks. Even when the flow might seem at first to be discontinuous, as when it drops precipitously over a cliff, a closer examination shows that it really is not. As the water approaches the cliff it speeds up. When it finally goes over it accelerates very quickly but no matter how fast it goes it moves continuously, moving from here to there by occupying every point in between. This is continuous motion. It never disappears over there and instantaneously reappears over here. That would be discontinuous motion.

Similarly, a thrown stone flies continuously (and smoothly) from release point to landing point, passing through each point in its path.

But wait.

If the stone passes through discrete points it must be doing so by teeny tiny little jerks, mustn't it? Otherwise how would it get from one point to the next? Is it possible that motion in the real world, much like motion in a movie, is really composed of tiny jerks from one point to the next but that these tiny jerks are simply too small and too fast for our senses to detect?

If so, then the real world is more like the rational number line ($\mathbb{Q}$) from Chapter 2 than the real number line ($\mathbb{R}$). In that case, motion really consists of jumping discretely over the "missing" points (like $\sqrt{2}$) as we move from here to there. That may seem like a bizarre idea to you — it does to us as well — but the idea of continuous motion is equally bizarre. It's just a little harder to see why.

The real world will be what it is regardless of what we believe it to be, but fortunately in mathematics we are *not* constrained to live in it. So we won't even try. We will simply postulate that no such jerkiness exists; that all motion is continuous.

However we *are* constrained to live with the logical consequences of our assumptions, once they are made. These will lead us into some very deep waters indeed.

The intuitive treatment of continuity was maintained throughout the 1700's as it was not generally perceived that a truly rigorous definition was necessary.

Consider the following definition given by Euler in 1748.

> A continuous curve is one such that its nature can be expressed by a single function of x. If a curve is of such a nature that for its various parts . . . different functions of x are required for its expression, . . . , then we call such a curve discontinuous.

However, the complexities associated with Fourier series and the types of functions that they represented caused mathematicians in the early 1800s to rethink their notions of continuity. As we saw in Part II, the graph of the function defined by the Fourier series

$$\frac{4}{\pi}\sum_{k=0}^{\infty}\frac{(-1)^k}{(2k+1)}\cos\left((2k+1)\,\pi x\right)$$

looked like this:

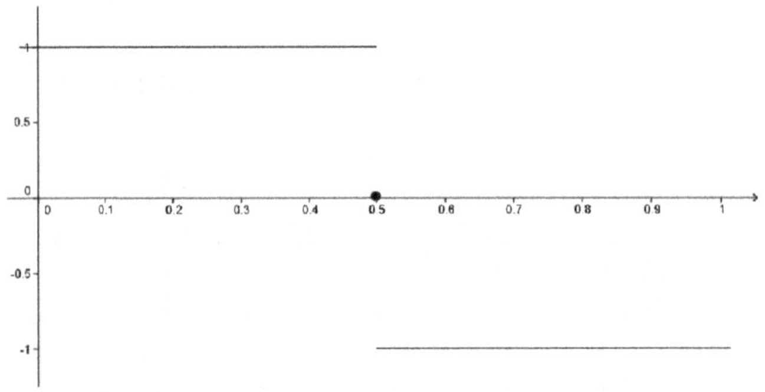

Figure 8.1.1

This function went against Euler's notion of what a continuous function should be. Here, an infinite sum of continuous cosine curves provided a single expression which resulted in a "discontinuous" curve. But as we've seen this didn't happen with power series and an intuitive notion of continuity is inadequate to explain the difference. Even more perplexing is the following situation. Intuitively, one would think that a continuous curve should have a tangent line at at least one point. It may have a number of jagged points to it, but it should be "smooth" somewhere. An example of this would be $f(x) = x^{2/3}$. Its graph is given by

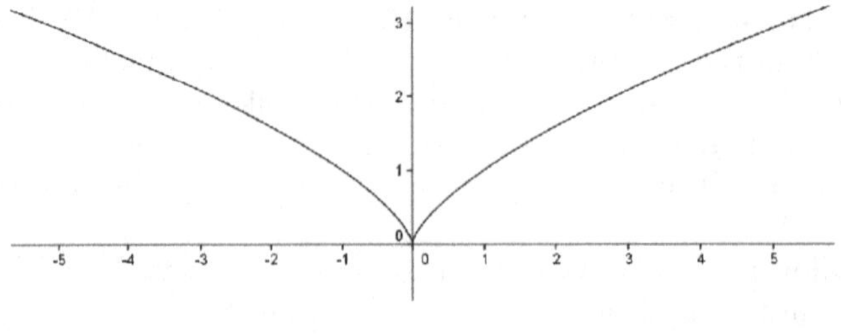

Figure 8.1.2

This function is not differentiable at the origin but it is differentiable everywhere else. One could certainly come up with examples of functions which fail to be differentiable at any number of points but, intuitively, it would be reasonable

Figure 8.1.3 Karl Weierstrass[15]

to expect that a continuous function should be differentiable *somewhere*. We might conjecture the following:

Conjecture 8.1.4 *If f is continuous on an interval I then there is some $a \in I$, such that $f'(a)$ exists.*

Surprisingly, in 1872, Karl Weierstrass showed that the above conjecture is ***FALSE***. He did this by displaying the counterexample:

$$f(x) = \sum_{n=0}^{\infty} b^n \cos(a^n \pi x).$$

Weierstrass showed that if a is an odd integer, $b \in (0,1)$, and $ab > 1 + \frac{3}{2}\pi$, then f is continuous everywhere, but is nowhere differentiable. Such a function is somewhat "fractal" in nature, and it is clear that a definition of continuity relying on intuition is inadequate to study it.

Problem 8.1.5

(a) Given $f(x) = \sum_{n=0}^{\infty} \left(\frac{1}{2}\right)^n \cos(a^n \pi x)$, what is the smallest value of a for which f satisfies Weierstrass' criterion to be continuous and nowhere differentiable.

(b) Let $f(x, N) = \sum_{n=0}^{N} \left(\frac{1}{2}\right)^n \cos(13^n \pi x)$ and use a computer algebra system to plot $f(x, N)$ for $N = 0, 1, 2, 3, 4, 10$ and $x \in [0, 1]$.

(c) Plot $f(x, 10)$ for $x \in [0, c]$, where $c = 0.1, 0.01, 0.001, 0.0001, 0.00001$. Based upon what you see in parts b and c, why would we describe the function to be somewhat "fractal" in nature?

□

Just as it was important to define convergence with a rigorous definition without appealing to intuition or geometric representations, it is imperative that we define continuity in a rigorous fashion not relying on graphs.

The first appearance of a definition of continuity which did not rely on geometry or intuition was given in 1817 by Bernhard Bolzano in a paper published in the Proceedings of the Prague Scientific Society entitled *Rein analytischer Beweis des Lehrsatzes dass zwieschen je zwey Werthen, die ein entgegengesetztes Resultat gewaehren, wenigstens eine reele Wurzel der Gleichung liege* (Purely Analytic Proof of the Theorem that Between Any Two Values that Yield Results of Opposite Sign There Will be at Least One Real Root of the Equation).

[15] mathshistory.st-andrews.ac.uk/Biographies/Weierstrass/

Figure 8.1.6 Bernhard Bolzano[16]

From the title it should be clear that in this paper Bolzano is proving the Intermediate Value Theorem. To do this he needs a completely analytic definition of continuity. The substance of Bolzano's idea is that if f is continuous at a point a then $f(x)$ should be "close to" $f(a)$ whenever x is "close enough to" a. More precisely, Bolzano said that f is continuous at a provided $|f(x) - f(a)|$ can be made smaller than any given quantity provided we make $|x - a|$ sufficiently small.

The language Bolzano uses is very similar to the language Leibniz used when he postulated the existence of infinitesimally small numbers. Leibniz said that infinitesimals are "smaller than any given quantity but not zero." Bolzano says that "$|f(x) - f(a)|$ can be made smaller than any given quantity provided we make $|x - a|$ sufficiently small." But Bolzano stops short of saying that $|x - a|$ is *infinitesimally* small. Given a, we can choose x so that $|x - a|$ is smaller than any real number we could name, say b, provided we name b *first*, but for any given choice of x, $|x - a|$, and b are both still real numbers. Possibly very small real numbers to be sure, but real numbers nonetheless. Infinitesimals have no place in Bolzano's construction.

Bolzano's paper was not well known when Cauchy proposed a similar definition in his *Cours d'analyse* [1] of 1821 so it is usually Cauchy who is credited with this definition, but even Cauchy's definition is not quite tight enough for modern standards. It was Karl Weierstrass in 1859 who finally gave the modern definition.

Definition 8.1.7 We say that a function $\boldsymbol{f}$ is continuous at $\boldsymbol{a}$ provided that for any $\varepsilon > 0$, there exists a $\delta > 0$ such that if $|x - a| < \delta$ then $|f(x) - f(a)| < \varepsilon$. ◊

Notice that the definition of continuity of a function is done point-by-point. A function can certainly be continuous at some points while discontinuous at others. When we say that f is continuous on an interval, then we mean that it is continuous at every point of that interval and, in theory, we would need to use the above definition to check continuity at each individual point.

Our definition fits the bill in that it does not rely on either intuition or graphs, but it is this very non-intuitiveness that makes it hard to grasp. It usually takes some time to become comfortable with this definition, let alone use it to prove theorems such as the Extreme Value Theorem and Intermediate Value Theorem. So let's go slowly to develop a feel for it.

This definition spells out a completely black and white procedure: you give me a positive number ε, and I must be able to find a positive number δ which satisfies a certain property. If I can always do that then the function is continuous at the point of interest.

[16] mathshistory.st-andrews.ac.uk/Biographies/Bolzano/

This definition also makes very precise what we mean when we say that $f(x)$ should be "close to" $f(a)$ whenever x is "close enough to" a. For example, intuitively we know that $f(x) = x^2$ should be continuous at $x = 2$. This means that we should be able to get x^2 to within, say, $\varepsilon = .1$ of 4 provided we make x close enough to 2. Specifically, we want $3.9 < x^2 < 4.1$. This happens exactly when $\sqrt{3.9} < x < \sqrt{4.1}$. Using the fact that $\sqrt{3.9} < 1.98$ and $2.02 < \sqrt{4.1}$, then we can see that if we get x to within $\delta = .02$ of 2, then $\sqrt{3.9} < 1.98 < x < 2.02 < \sqrt{4.1}$ and so x^2 will be within .1 of 4. This is very straightforward. What makes this situation more difficult is that we must be able to do this for any $\varepsilon > 0$.

Notice the similarity between this definition and the definition of convergence of a sequence. Both definitions have the challenge of an $\varepsilon > 0$. In the definition of $\lim_{n\to\infty} s_n = s$, we had to get s_n to within ε of s by making n large enough. For sequences, the challenge lies in making $|s_n - s|$ sufficiently small. More precisely, given $\varepsilon > 0$ we need to decide how large n should be to guarantee that $|s_n - s| < \varepsilon$.

In our definition of continuity, we still need to make something small (namely $|f(x) - f(a)| < \varepsilon$), only this time, we need to determine how close x must be to a to ensure this will happen instead of determining how large n must be.

What makes f continuous at a is the arbitrary nature of ε (as long as it is positive). As ε becomes smaller, this forces $f(x)$ to be closer to $f(a)$. That we can always find a positive distance δ to work is what we mean when we say that we can make $f(x)$ as close to $f(a)$ as we wish, provided we get x close enough to a. The sequence of pictures below illustrates that the phrase "for any $\varepsilon > 0$, there exists a $\delta > 0$ such that if $|x - a| < \delta$ then $|f(x) - f(a)| < \varepsilon$" can be replaced by the equivalent formulation "for any $\varepsilon > 0$, there exists a $\delta > 0$ such that if $a - \delta < x < a + \delta$ then $f(a) - \varepsilon < f(x) < f(a) + \varepsilon$." This could also be replaced by the phrase "for any $\varepsilon > 0$, there exists a $\delta > 0$ such that if $x \in (a - \delta, a + \delta)$ then $f(x) \in (f(a) - \varepsilon, f(a) + \varepsilon)$." All of these equivalent formulations convey the idea that we can get $f(x)$ to within ε of $f(a)$, provided we make x within δ of a, and we will use whichever formulation suits our needs in a particular application.

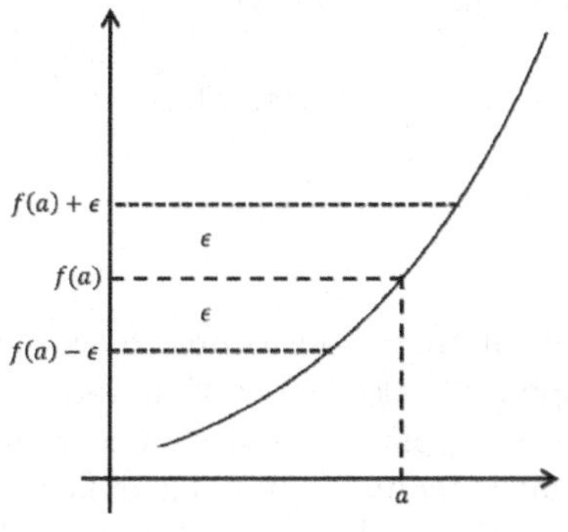

Given a distance ϵ away from $f(a)$

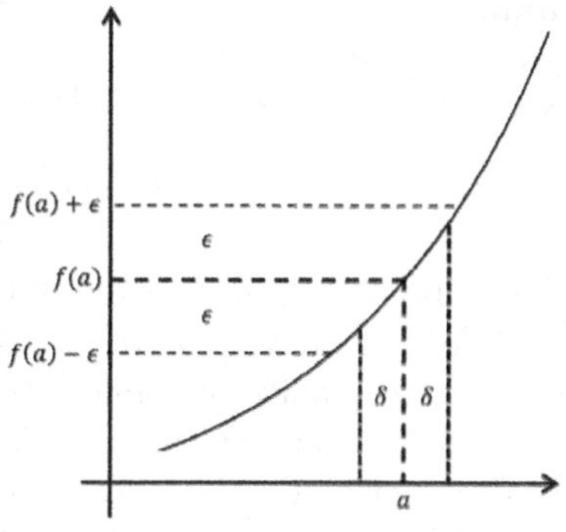

we must find a distance δ away from a

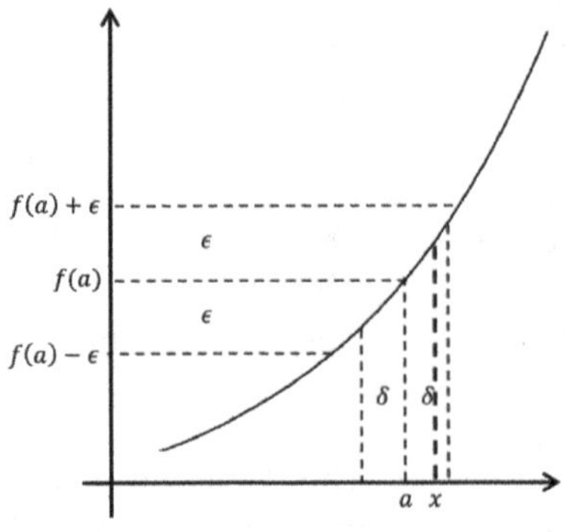

such that whenever $a - \delta < x < a + \delta$

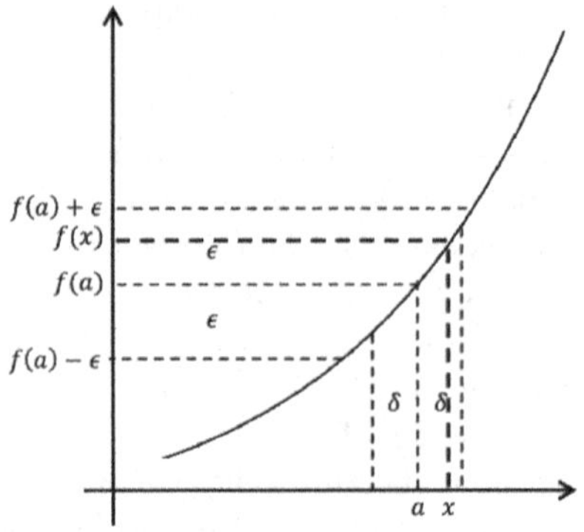

$f(a) - \varepsilon < f(x) < f(a) + \varepsilon$

The precision of the definition is what allows us to examine continuity without relying on pictures or vague notions such as "nearness" or "getting closer to." We will now consider some examples to illustrate this precision.

Example 8.1.8 Use the definition of continuity to show that $f(x) = x$ is continuous at any point a. □

If we were to draw the graph of this line, then you would likely say that this is obvious. The point behind the definition is that we can back up your intuition in a rigorous manner.

Proof. Let $\varepsilon > 0$. Let $\delta = \varepsilon$. If $|x - a| < \delta$, then

$$|f(x) - f(a)| = |x - a| < \varepsilon$$

Thus by the definition, f is continuous at a. ■

Problem 8.1.9 Use the definition of continuity to show that if m and b are fixed (but unspecified) real numbers then the function

$$f(x) = mx + b$$

is continuous at every real number a. □

Example 8.1.10 Use the definition of continuity to show that $f(x) = x^2$ is continuous at $a = 0$. □

Proof. Let $\varepsilon > 0$. Let $\delta = \sqrt{\varepsilon}$. If $|x - 0| < \delta$, then $|x| < \sqrt{\varepsilon}$. Thus

$$\left|x^2 - 0^2\right| = |x|^2 < \left(\sqrt{\varepsilon}\right)^2 = \varepsilon.$$

Thus by the definition, f is continuous at 0. ■

Notice that in these proofs, the challenge of an $\varepsilon > 0$ was first given. This is because the choice of δ must depend upon ε. Also notice that there was no explanation for our choice of δ. We just supplied it and showed that it worked. As long as $\delta > 0$, then this is all that is required. In point of fact, the δ we chose in each example was not the only choice that worked; any smaller δ would work as well.

Problem 8.1.11

(a) Given a particular $\varepsilon > 0$ in the definition of continuity, show that if a particular $\delta_0 > 0$ satisfies the definition, then any δ with $0 < \delta < \delta_0$ will also work for this ε.

(b) Show that if a δ can be found to satisfy the conditions of the definition of continuity for a particular $\varepsilon_0 > 0$, then this δ will also work for any ε

with $0 < \varepsilon_0 < \varepsilon$.

□

It wasn't explicitly stated in the definition but when we say "if $|x - a| < \delta$ then $|f(x) - f(a)| < \varepsilon$," we should be restricting ourselves to x values which are in the domain of the function f, otherwise $f(x)$ doesn't make sense. We didn't put it in the definition because that definition was complicated enough without this technicality. Also in the above examples, the functions were defined everywhere so this was a moot point. We will continue with the convention that when we say "if $|x - a| < \delta$ then $|f(x) - f(a)| < \varepsilon$," we will be restricting ourselves to x values which are in the domain of the function f. This will allow us to examine continuity of functions not defined for all x without restating this restriction each time.

Problem 8.1.12 Use the definition of continuity to show that

$$f(x) = \begin{cases} \sqrt{x} & \text{if } x \geq 0 \\ -\sqrt{-x} & \text{if } x < 0 \end{cases}$$

is continuous at $a = 0$. □

Problem 8.1.13 Use the definition of continuity to show that $f(x) = \sqrt{x}$ is continuous at $a = 0$. How is this problem different from problem 8.1.12? How is it similar? □

Sometimes the δ that will work for a particular ε is fairly obvious to see, especially after you've gained some experience. This is the case in the above examples (at least after looking back at the proofs). However, the task of finding a δ to work is usually not so obvious and requires some scrapwork. This scrapwork is vital toward producing a δ, but again is not part of the polished proof. This can be seen in the following example.

Example 8.1.14 Use the definition of continuity to prove that $f(x) = \sqrt{x}$ is continuous at $a = 1$.

SCRAPWORK

As before, the scrapwork for these problems often consists of simply working backwards. Specifically, given an $\varepsilon > 0$, we need to find a $\delta > 0$ so that $|\sqrt{x} - \sqrt{1}| < \varepsilon$, whenever $|x - 1| < \delta$. We work backwards from what we want, keeping an eye on the fact that we can control the size of $|x - 1|$.

$$|\sqrt{x} - \sqrt{1}| = |\frac{(\sqrt{x} - 1)(\sqrt{x} + 1)}{\sqrt{x} + 1}| = \frac{|x - 1|}{\sqrt{x} + 1} < |x - 1|.$$

This seems to suggest that we should make $\delta = \varepsilon$. We're now ready for the formal proof. □

Proof. Let $\varepsilon > 0$. Let $\delta = \varepsilon$. If $|x - 1| < \delta$, then $|x - 1| < \varepsilon$, and so

$$\left|\sqrt{x} - \sqrt{1}\right| = |\frac{(\sqrt{x} - 1)(\sqrt{x} + 1)}{\sqrt{x} + 1}| = \frac{|x - 1|}{\sqrt{x} + 1} < |x - 1| < \varepsilon.$$

Thus by definition, $f(x) = \sqrt{x}$ is continuous at 1. ■

Figure 8.1.15 Paul Halmos[17]

Bear in mind that someone reading the formal proof will not have seen the scrapwork, so the choice of δ might seem rather mysterious. However, you are in no way bound to motivate this choice of δ and usually you should not, unless it is necessary for the formal proof. All you have to do is find this δ and show that it works. Furthermore, to a trained reader, your ideas will come through when you demonstrate that your choice of δ works.

Now reverse this last statement. *As* a trained reader, when you read the proof of a theorem it is *your* responsibility to find the scrapwork, to see how the proof works and understand it fully. As the renowned mathematical expositor Paul Halmos (1916-2006) said, "Don't just read it; fight it! Ask your own questions, look for your own examples, discover your own proofs. Is the hypothesis necessary? Is the converse true? What happens in the classical special case? What about the degenerate cases? Where does the proof use the hypothesis?"

This is the way to learn mathematics. It is really the only way.

Problem 8.1.16 Use the definition of continuity to show that $f(x) = \sqrt{x}$ is continuous at any positive real number a. □

Problem 8.1.17

(a) Use a unit circle to show that for $0 \leq \theta < \frac{\pi}{2}$, $\sin\theta \leq \theta$ and $1 - \cos\theta \leq \theta$ and conclude $|\sin\theta| \leq |\theta|$ and $|1 - \cos\theta| \leq |\theta|$ for $-\frac{\pi}{2} < \theta < \frac{\pi}{2}$.

(b) Use the definition of continuity to prove that $f(x) = \sin x$ is continuous at any point a.

Hint.

$\sin x = \sin(x - a + a)$.

□

Problem 8.1.18

(a) Use the definition of continuity to show that $f(x) = e^x$ is continuous at $a = 0$.

(b) Show that $f(x) = e^x$ is continuous at any point a.

Hint.

Rewrite $e^x - e^a$ as $e^{a+(x-a)} - e^a$ and use what you proved in part a.

□

[17] mathshistory.st-andrews.ac.uk/Biographies/Halmos/

In the above problems, we used the definition of continuity to verify our intuition about the continuity of familiar functions. The advantage of this analytic definition is that it can be applied when the function is not so intuitive. Consider, for example, the function given at the end of the last chapter.

$$f(x) = \begin{cases} x \sin\left(\frac{1}{x}\right), & \text{if } x \neq 0 \\ 0, & \text{if } x = 0 \end{cases}.$$

Near zero, the graph of $f(x)$ looks like this:

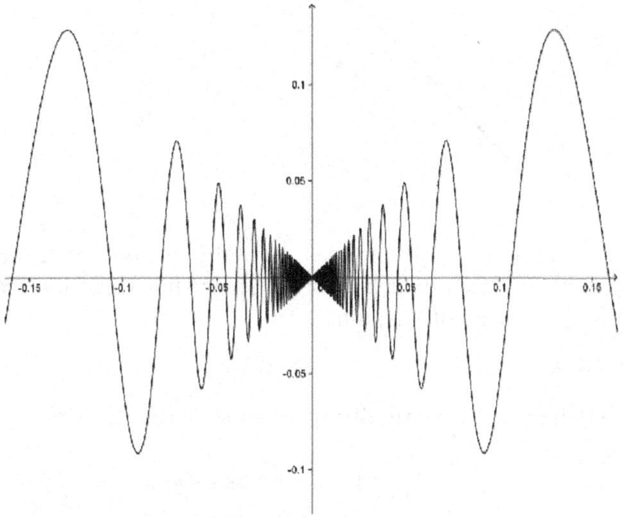

As we mentioned in the previous chapter, since $\sin\left(\frac{1}{x}\right)$ oscillates infinitely often as x nears zero this graph must be viewed with a certain amount of suspicion. However our completely analytic definition of continuity shows that this function is, in fact, continuous at 0.

Problem 8.1.19 The Topologist's Sine Function. Use the definition of continuity to show that

$$f(x) = \begin{cases} x \sin\left(\frac{1}{x}\right), & \text{if } x \neq 0 \\ 0, & \text{if } x = 0 \end{cases}$$

is continuous at 0. □

Even more perplexing is the function defined by

$$D(x) = \begin{cases} x, & \text{if } x \text{ is rational} \\ 0, & \text{if } x \text{ is irrational.} \end{cases}$$

To the naked eye, the graph of this function looks like the lines $y = 0$ and $y = x$. Of course, such a graph would not be the graph of a function. Actually, both of these lines have holes in them. Wherever there is a point on one line there is a "hole" on the other. Each of these holes is the width of a single point (that is, their "width" is zero!) so they are invisible to the naked eye (or even magnified under the most powerful microscope available). This idea is illustrated in the following graph

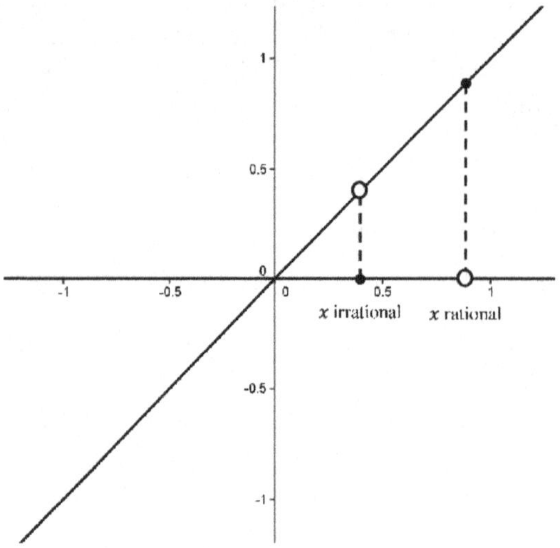

Can such a function so "full of holes" actually be continuous anywhere? It turns out that we can use our definition to show that this function is, in fact, continuous at 0 and at no other point.

Problem 8.1.20

(a) Use the definition of continuity to show that the function

$$D(x) = \begin{cases} x, & \text{if } x \text{ is rational} \\ 0, & \text{if } x \text{ is irrational} \end{cases}$$

is continuous at 0.

(b) Let $a \neq 0$. Use the definition of continuity to show that D is not continuous at a.

Hint.

You might want to break this up into two cases where a is rational or irrational. Show that no choice of $\delta > 0$ will work for $\varepsilon = |a|$. Note that Theorem 2.0.11 of Chapter 2 will probably help here.

□

8.2 Sequences and Continuity

There is an alternative way to prove that the function

$$D(x) = \begin{cases} x, & \text{if } x \text{ is rational} \\ 0, & \text{if } x \text{ is irrational} \end{cases}$$

is not continuous at $a \neq 0$. We will examine this by looking at the relationship between our definitions of convergence and continuity. The two ideas are actually quite closely connected, as illustrated by the following very useful theorem.

Theorem 8.2.1 *The function f is continuous at a if and only if f satisfies the following property:*

$$\forall \text{ sequences } (x_n),\ \text{if } \lim_{n\to\infty} x_n = a \text{ then } \lim_{n\to\infty} f(x_n) = f(a).$$

Theorem 8.2.1 says that in order for f to be continuous, it is necessary and sufficient that any sequence (x_n) converging to a must force the sequence $(f(x_n))$ to converge to $f(a)$. A picture of this situation is below though, as always, the formal proof will not rely on the diagram.

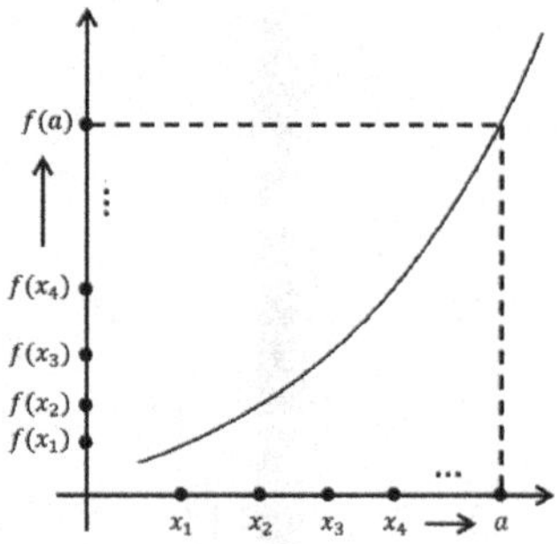

This theorem is especially useful for showing that a function f, is not continuous at a point a; all we need to do is exhibit a sequence (x_n) converging to a such that the sequence $\lim_{n\to\infty} f(x_n)$ does *not* converge to $f(a)$. Let's demonstrate this idea before we tackle the proof of Theorem 8.2.1.

Example 8.2.2 Use Theorem 8.2.1 to prove that

$$f(x) = \begin{cases} \frac{|x|}{x}, & \text{if } x \neq 0 \\ 0, & \text{if } x = 0 \end{cases}$$

is not continuous at 0. □

Proof. First notice that f can be written as

$$f(x) = \begin{cases} 1 & \text{if } x > 0 \\ -1 & \text{if } x < 0 \\ 0 & \text{if } x = 0 \end{cases}.$$

To show that f is not continuous at 0, all we need to do is create a single sequence (x_n)which converges to 0, but for which the sequence $(f(x_n))$ does not converge to $f(0) = 0$. For a function like this one, just about any sequence will do, but let's use $\left(\frac{1}{n}\right)$, just because it is an old familiar friend.

We have $\lim\limits_{n\to\infty} \frac{1}{n} = 0$, but $\lim\limits_{n\to\infty} f\left(\frac{1}{n}\right) = \lim\limits_{n\to\infty} 1 = 1 \neq 0 = f(0)$. Thus by Theorem 8.2.1, f is not continuous at 0. ■

Problem 8.2.3 Use Theorem 8.2.1 to show that

$$f(x) = \begin{cases} \frac{|x|}{x}, & \text{if } x \neq 0 \\ a, & \text{if } x = 0 \end{cases}$$

is not continuous at 0, no matter what value a is. □

Problem 8.2.4 Use Theorem 8.2.1 to show that

$$D(x) = \begin{cases} x, & \text{if } x \text{ is rational} \\ 0, & \text{if } x \text{ is irrational} \end{cases}$$

is not continuous at $a \neq 0$. □

Problem 8.2.5 The function $T(x) = \sin\left(\frac{1}{x}\right)$ is often called the topologist's sine curve. Whereas $\sin x$ has roots at $n\pi$, $n \in \mathbb{Z}$ and oscillates infinitely often as $x \to \pm\infty$, T has roots at $\frac{1}{n\pi}$, $n \in \mathbb{Z}$, $n \neq 0$, and oscillates infinitely often as x approaches zero. A rendition of the graph follows.

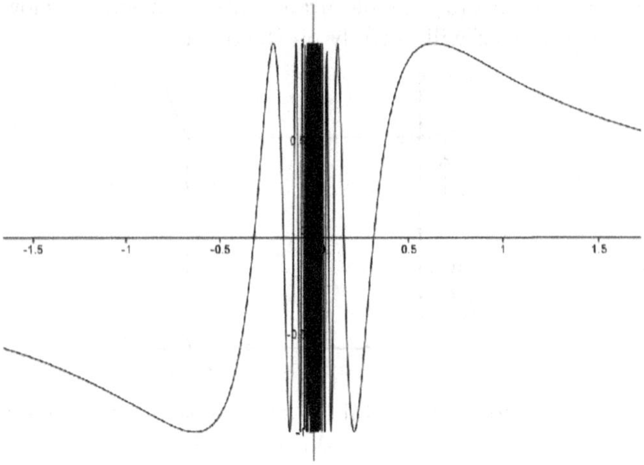

Notice that T is not even defined at $x = 0$. We can extend T to be defined at 0 by simply choosing a value for $T(0)$:

$$T(x) = \begin{cases} \sin\left(\frac{1}{x}\right), & \text{if } x \neq 0 \\ b, & \text{if } x = 0 \end{cases}.$$

Use Theorem 8.2.1 to show that T is not continuous at 0, no matter what value is chosen for b. □

Sketch of the Proof of Theorem 8.2.1. We've seen how we can use Theorem 8.2.1, now we need to prove Theorem 8.2.1. The forward direction is fairly straightforward. So we assume that f is continuous at a and start with a sequence (x_n) which converges to a. What is left to show is that $\lim_{n\to\infty} f(x_n) = f(a)$. If you write down the definitions of f being continuous at a, $\lim_{n\to\infty} x_n = a$, and $\lim_{n\to\infty} f(x_n) = f(a)$, you should be able to get from what you are assuming to what you want to conclude.

To prove the converse, it is convenient to prove its contrapositive. That is, we want to prove that if f is not continuous at a then we can construct a sequence (x_n) that converges to a but $(f(x_n))$does not converge to $f(a)$. First we need to recognize what it means for f to not be continuous at a. This says that somewhere there exists an $\varepsilon > 0$, such that no choice of $\delta > 0$ will work for this. That is, for any such δ, there will exist x, such that $|x - a| < \delta$, but $|f(x) - f(a)| \geq \varepsilon$. With this in mind, if $\delta = 1$, then there will exist an x_1 such that $|x_1 - a| < 1$, but $|f(x_1) - f(a)| \geq \varepsilon$. Similarly, if $\delta = \frac{1}{2}$, then there will exist an x_2 such that $|x_2 - a| < \frac{1}{2}$, but $|f(x_2) - f(a)| \geq \varepsilon$. If we continue in this fashion, we will create a sequence (x_n) such that $|x_n - a| < \frac{1}{n}$, but $|f(x_n) - f(a)| \geq \varepsilon$. This should do the trick. ■

Problem 8.2.6 Turn the ideas of the previous two paragraphs into a formal proof of Theorem 8.2.1. □

Theorem 8.2.1 is a very useful result. It is a bridge between the ideas of convergence and continuity so it allows us to bring all of the theory we developed in Chapter 6 to bear on continuity questions. For example consider the following.

Theorem 8.2.7 *Suppose f and g are both continuous at a. Then $f + g$ and $f \cdot g$ are continuous at a.*

Proof. We could use the definition of continuity to prove Theorem 8.2.7, but Theorem 8.2.1 makes our job much easier. For example, to show that $f + g$ is continuous, consider any sequence (x_n) which converges to a. Since f is continuous at a, then by Theorem 8.2.1, $\lim_{n\to\infty} f(x_n) = f(a)$. Likewise, since g is continuous at a, then $\lim_{n\to\infty} g(x_n) = g(a)$.
By Theorem 6.2.4 of Chapter 6,

$$\begin{aligned}\lim_{n\to\infty}(f+g)(x_n) &= \lim_{n\to\infty}(f(x_n)+g(x_n))\\ &= \lim_{n\to\infty} f(x_n) + \lim_{n\to\infty} g(x_n)\\ &= f(a)+g(a)\\ &= (f+g)(a).\end{aligned}$$

Thus by Theorem 8.2.1, $f + g$ is continuous at a. The proof that $f \cdot g$ is continuous at a is similar. ■

Problem 8.2.8 Use Theorem 8.2.1 to show that if f and g are continuous at a, then $f \cdot g$ is continuous at a. □

By employing Theorem 8.2.7 a finite number of times, we can see that a finite sum of continuous functions is continuous. That is, if f_1, f_2, ..., f_n are all continuous at a then $\sum_{j=1}^{n} f_j$ is continuous at a. But what about an infinite sum? Specifically, suppose f_1, $f_2, f_3, \ldots$ are all continuous at a. Consider the following argument.

Let $\varepsilon > 0$. Since f_j is continuous at a, then there exists $\delta_j > 0$ such that if $|\,x - a| < \delta_j$, then $|f_j(x) - f_j(a)| < \frac{\varepsilon}{2^j}$. Let $\delta = \min(\delta_1, \delta_2, \ldots)$. If $|\,x - a| < \delta$, then

$$\begin{aligned}\left|\sum_{j=1}^{\infty} f_j(x) - \sum_{j=1}^{\infty} f_j(a)\right| &= \left|\sum_{j=1}^{\infty}(f_j(x)-f_j(a))\right|\\ &\le \sum_{j=1}^{\infty}|f_j(x)-f_j(a)| < \sum_{j=1}^{\infty}\frac{\varepsilon}{2^j} = \varepsilon.\end{aligned}$$

Thus by definition, $\sum_{j=1}^{\infty} f_j$ is continuous at a.

This argument seems to say that an infinite sum of continuous functions must be continuous (provided it converges). However we know that the Fourier series

$$\frac{4}{\pi}\sum_{k=0}^{\infty}\frac{(-1)^k}{(2k+1)}\cos\left((2k+1)\,\pi x\right)$$

is a counterexample to this, as it is an infinite sum of continuous functions which does not converge to a continuous function. Something fundamental seems to have gone wrong here. Can you tell what it is?

This is a question we will spend considerable time addressing in Chapter 10 (in particular, see problem 10.1.1) so if you don't see the difficulty, don't worry; you will. In the meantime keep this problem tucked away in your consciousness. It is, as we said, fundamental.

Theorem 8.2.1 will also handle quotients of continuous functions. There is however a small detail that needs to be addressed first. Obviously, when we consider the continuity of f/g at a,we need to assume that $g(a) \neq 0$. However, g may be zero at other values. How do we know that when we choose our sequence (x_n) converging to a that $g(x_n)$ is not zero? This would mess up our idea of using the corresponding theorem for sequences (Theorem 6.2.12 from Chapter 6). This can be handled with the following lemma.

Lemma 8.2.9 *If g is continuous at a and $g(a) \neq 0$, then there exists $\delta > 0$ such that $g(x) \neq 0$ for all $x \in (a - \delta, a + \delta)$.*

Problem 8.2.10 Prove Lemma 8.2.9.

Hint.

Consider the case where $g(a) > 0$. Use the definition with $\varepsilon = \frac{g(a)}{2}$. The picture is below; make it formal.

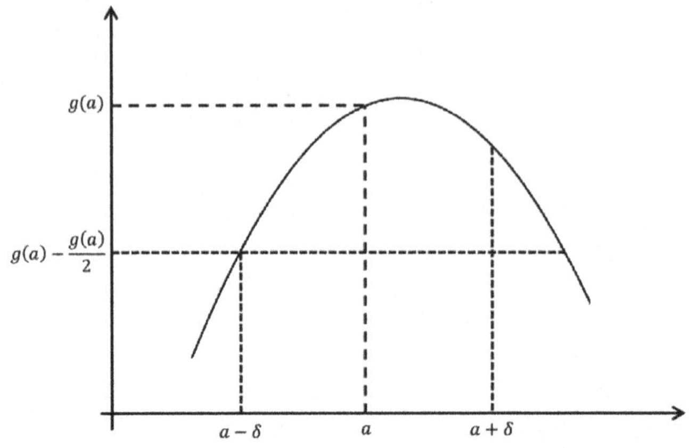

For the case $g(a) < 0$, consider the function $-g$. □

A consequence of this lemma is that if we start with a sequence (x_n) converging to a, then for n sufficiently large, $g(x_n) \neq 0$.

Problem 8.2.11 Use Theorem 8.2.1, to prove that if f and g are continuous at a and $g(a) \neq 0$, then f/g is continuous at a. □

Theorem 8.2.12 *Suppose f is continuous at a and g is continuous at $f(a)$. Then $g \circ f$ is continuous at a. (Note that $(g \circ f)(x) = g(f(x))$.)*

Problem 8.2.13 Prove Theorem 8.2.12

(a) Using the definition of continuity.

(b) Using Theorem 8.2.1.

□

The above theorems allow us to build continuous functions from other continuous functions. For example, knowing that $f(x) = x$ and $g(x) = c$ are continuous, we can conclude that any polynomial,

$$p(x) = a_n x^n + a_{n-1} x^{n-1} + \cdots + a_1 x + a_0$$

is continuous as well. We also know that functions such as $f(x) = \sin(e^x)$ are continuous without having to rely on the definition.

Problem 8.2.14 Show that each of the following is a continuous function at every point in its domain.

(a) Any polynomial.

(b) Any rational function. (A rational function is defined to be a ratio of polynomials.)

(c) $\cos x$.

(d) The other trig functions: $\tan(x)$, $\cot(x)$, $\sec(x)$, and $\csc(x)$.

□

Problem 8.2.15 What allows us to conclude that $f(x) = \sin(e^x)$ is continuous at any point a without referring back to the definition of continuity? □

Theorem 8.2.1 can also be used to study the convergence of sequences. For example, since $f(x) = e^x$ is continuous at any point and $\lim_{n\to\infty} \frac{n+1}{n} = 1$, then $\lim_{n\to\infty} e^{\left(\frac{n+1}{n}\right)} = e$. This also illustrates a certain way of thinking about continuous functions. They are the ones where we can "commute" the function and a limit of a sequence. Specifically, if f is continuous at a and $\lim_{n\to\infty} x_n = a$, then $\lim_{n\to\infty} f(x_n) = f(a) = f(\lim_{n\to\infty} x_n)$.

Problem 8.2.16 Compute the following limits. Be sure to point out how continuity is involved.

(a) $\displaystyle\lim_{n\to\infty} \sin\left(\frac{n\pi}{2n+1}\right)$

(b) $\displaystyle\lim_{n\to\infty} \sqrt{\frac{n}{n^2+1}}$

(c) $\displaystyle\lim_{n\to\infty} e^{(\sin(1/n))}$

□

Having this rigorous formulation of continuity is necessary for proving the Extreme Value Theorem and the Mean Value Theorem. However there is one more piece of the puzzle to address before we can prove these theorems.

We will do this in the next chapter, but before we go on it is time to define a fundamental concept that was probably one of the first you learned in calculus: limits.

8.3 The Definition of the Limit of a Function

Since these days the limit concept is generally regarded as the starting point for calculus, you might think it is a little strange that we've chosen to talk about continuity first. But historically, the formal definition of a limit came after the formal definition of continuity. In some ways, the limit concept was part of a unification of all the ideas of calculus that were studied previously and, subsequently, it became the basis for all ideas in calculus. For this reason it is logical to make it the first topic covered in a calculus course.

To be sure, limits were always lurking in the background. In his attempts to justify his calculations, Newton used what he called his doctrine of "Ultimate Ratios." Specifically the ratio $\frac{(x+h)^2-x^2}{h} = \frac{2xh+h^2}{h} = 2x + h$ becomes the ultimate ratio $2x$ at the last instant of time before h - an "evanescent quantity" - vanishes ([4], p. 33). Similarly Leibniz's "infinitely small" differentials dx and dy can be seen as an attempt to get "arbitrarily close" to x and y, respectively. This is the idea at the heart of calculus: to get arbitrarily close to, say, x without actually reaching it.

As we saw in Chapter 4, Lagrange tried to avoid the entire issue of "arbitrary closesness," both in the limit and differential forms when, in 1797, he attempted to found calculus on infinite series.

Although Lagrange's efforts failed, they set the stage for Cauchy to provide a definition of derivative which in turn relied on his precise formulation of a limit. Consider the following example: to determine the slope of the tangent line (derivative) of $f(x) = \sin x$ at $x = 0$. We consider the graph of the difference quotient $D(x) = \frac{\sin x}{x}$.

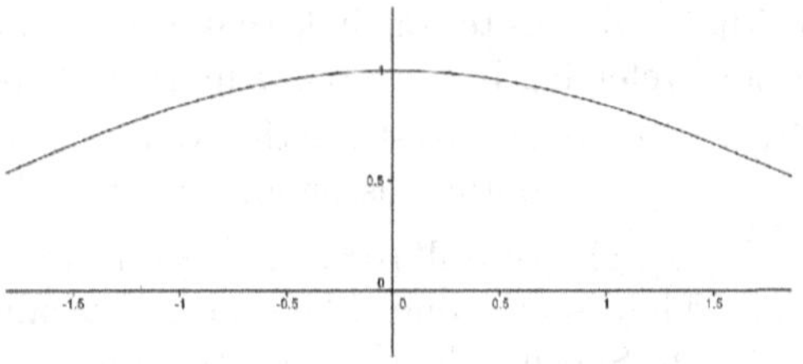

From the graph, it appears that $D(0) = 1$ but we must be careful. $D(0)$ doesn't even exist! Somehow we must convey the idea that $D(x)$ will approach 1 as x approaches 0, even though the function is not defined at 0. Cauchy's idea was that the limit of $D(x)$ would equal 1 because we can make $D(x)$ differ from 1 by as little as we wish ([6], p. 158).

Karl Weierstrass made these ideas precise in his lectures on analysis at the University of Berlin (1859-60) and provided us with our modern formulation.

Definition 8.3.1 We say $\lim_{x \to a} f(x) = L$ provided that for each $\varepsilon > 0$, there exists $\delta > 0$ such that if $0 < |x - a| < \delta$ then $|f(x) - L| < \varepsilon$. ◊

Before we delve into this, notice that it is very similar to the definition of the continuity of $f(x)$ at $x = a$. In fact we can readily see that f is continuous at $x = a$ if and only if $\lim_{x \to a} f(x) = f(a)$.

Comment. This presumes that a is an **accumulation point** of the domain of f (Definition 11.0.5). We will discuss accumulation points in Chapter 11.

There are two differences between this definition and the definition of continuity and they are related. The first is that we replace the value $f(a)$ with L. This is because the function may not be defined at a. In a sense the limiting value L is the value f would have *if it were defined and continuous at* a. The second is that we have replaced

$$|x - a| < \delta$$

with

$$0 < |x - a| < \delta.$$

Again, since f needn't be defined at a, we will not even consider what happens when $x = a$. This is the only purpose for this change.

As with the definition of the limit of a sequence, this definition does not determine what L is, it only verifies that your guess for the value of the limit is correct.

Finally, a few comments on the differences and similiarities between this limit and the limit of a sequence are in order, if for no other reason than because we use the same notation (lim) for both.

When we were working with sequences in Chapter 6 and wrote things like $\lim_{n \to \infty} a_n$ we were thinking of n as an integer that got bigger and bigger. To put that more mathematically, the limit parameter n was taken from the set of positive integers, or $n \in \mathbb{N}$.

For both continuity and the limit of a function we write things like $\lim_{x \to a} f(x)$ and think of x as a variable that gets arbitrarily close to the number a. Again, to be more mathematical in our language we would say that the limit parameter x is taken from the ... Well, actually, this is interesting isn't it? Do we need to take x from $\mathbb{Q}$ or from $\mathbb{R}$? The requirement in both cases is simply that we be able to choose x arbitrarily close to a. From Theorem 2.0.11 of Chapter 2 we see that this is possible whether x is rational or not, so it seems either will work. This leads to the pardoxical sounding conclusion that we do not need a continuum ($\mathbb{R}$) to have continuity. This seems strange.

Before we look at the above example, let's look at some algebraic examples to see the definition in use.

Example 8.3.2 Consider the function $D(x) = \frac{x^2-1}{x-1}$, $x \neq 1$. You probably recognize this as the difference quotient used to compute the derivative of $f(x) = x^2$ at $x = 1$, so we strongly suspect that $\lim\limits_{x\to 1} \dfrac{x^2-1}{x-1} = 2$. Just as when we were dealing with limits of sequences, we should be able to use the definition to verify this. And as before, we will start with some scrapwork.

SCRAPWORK

Let $\varepsilon > 0$. We wish to find a $\delta > 0$ such that if $0 < |x-1| < \delta$ then $\left|\frac{x^2-1}{x-1} - 2\right| < \varepsilon$. With this in mind, we perform the following calculations

$$\left|\frac{x^2-1}{x-1} - 2\right| = |(x+1) - 2| = |x-1|.$$

Now we have a handle on δ that will work in the definition and we'll give the formal proof that

$$\lim_{x\to 1} \frac{x^2-1}{x-1} = 2.$$

□

Proof. Let $\varepsilon > 0$ and let $\delta = \varepsilon$. If $0 < |x-1| < \delta$, then

$$\left|\frac{x^2-1}{x-1} - 2\right| = |(x+1) - 2| = |x-1| < \delta = \varepsilon.$$

■

As in our previous work with sequences and continuity, notice that the scrapwork is not part of the formal proof (though it was necessary to determine an appropriate δ). Also, notice that $0 < |x-1|$ was not really used except to ensure that $x \neq 1$.

Problem 8.3.3 Use the definition of a limit to verify that

$$\lim_{x\to a} \frac{x^2-a^2}{x-a} = 2a.$$

□

Problem 8.3.4 Use the definition of a limit to verify each of the following limits.

(a) $\lim\limits_{x\to 1} \dfrac{x^3-1}{x-1} = 3$

Hint.

$$\begin{aligned}\left|\frac{x^3-1}{x-1} - 3\right| &= \left|x^2+x+1-3\right| \\ &\leq \left|x^2-1\right| + |x-1| \\ &= \left|(x-1+1)^2 - 1\right| + |x-1| \\ &= \left|(x-1)^2 + 2(x-1)\right| + |x-1| \\ &\leq |x-1|^2 + 3\,|x-1|.\end{aligned}$$

(b) $\lim_{x\to 1} \frac{\sqrt{x}-1}{x-1} = 1/2$

Hint.

$$\begin{aligned} \left|\frac{\sqrt{x}-1}{x-1} - \frac{1}{2}\right| &= \left|\frac{1}{\sqrt{x}+1} - \frac{1}{2}\right| \\ &= \left|\frac{2-(\sqrt{x}+1)}{2(\sqrt{x}+1)}\right| \\ &= \left|\frac{1-x}{2(1+\sqrt{x})^2}\right| \\ &\le \frac{1}{2}|x-1|. \end{aligned}$$

□

Let's go back to the original problem: to show that $\lim_{x\to 0} \frac{\sin x}{x} = 1$.

While rigorous, our definition of continuity is quite cumbersome. We really need to develop some tools we can use to show continuity rigorously without having to refer directly to the definition. We have already seen in Theorem 8.2.1 one way to do this. Here is another. The key is the observation we made after the definition of a limit:

$$f \text{ is continuous at } x = a \text{ if and only if } \lim_{x\to a} f(x) = f(a).$$

Read another way, we could say that $\lim_{x\to a} f(x) = L$ provided that if we redefine $f(a) = L$ (or define $f(a) = L$ in the case where $f(a)$ is not defined) then f becomes continuous at a. This allows us to use all of the machinery we proved about continuous functions and limits of sequences.

For example, the following corollary to Theorem 8.2.1 comes virtually for free once we've made the observation above.

Corollary 8.3.5 $\lim_{x\to a} f(x) = L$ *if and only if f satisfies the following property:*

$$\forall \text{ sequences } (x_n), x_n \neq a, \text{ if } \lim_{n\to\infty} x_n = a \text{ then } \lim_{n\to\infty} f(x_n) = L.$$

Armed with this, we can prove the following familiar limit theorems from calculus.

Theorem 8.3.6 *Suppose* $\lim_{x\to a} f(x) = L$ *and* $\lim_{x\to a} g(x) = M$*, then*

a $\lim_{x\to a} (f(x) + g(x)) = L + M$

b $\lim_{x\to a} (f(x) \cdot g(x)) = L \cdot M$

c $\lim_{x\to a} \left(\frac{f(x)}{g(x)}\right) = L/M$ *provided $M \neq 0$ and $g(x) \neq 0$, for x sufficiently close to a (but not equal to a).*

We will prove part (a) to give you a feel for this and let you prove parts (b) and (c).

Proof. Let (x_n) be a sequence such that $x_n \neq a$ and $\lim_{n\to\infty} x_n = a$. Since $\lim_{x\to a} f(x) = L$ and $\lim_{x\to a} g(x) = M$ we see that $\lim_{n\to\infty} f(x_n) = L$ and $\lim_{n\to\infty} g(x_n) = M$. By Theorem 6.2.4 of Chapter 6, we have $\lim_{n\to\infty} f(x_n) + g(x_n) = L + M$.

Since $\{x_n\}$ was an arbitrary sequence with $x_n \neq a$ and $\lim_{n\to\infty} x_n = a$ we have

$$\lim_{x\to a} f(x) + g(x) = L + M.$$

■

Problem 8.3.7 Prove parts (b) and (c) of Theorem 8.3.6. □

More in line with our current needs, we have a reformulation of the Squeeze Theorem.

Theorem 8.3.8 ***Squeeze Theorem for functions***

Suppose $f(x) \leq g(x) \leq h(x)$, for x sufficiently close to a (but not equal to a). If $\lim_{x\to a} f(x) = L = \lim_{x\to a} h(x)$, then $\lim_{x\to a} g(x) = L$ also.

Problem 8.3.9 Prove Theorem 8.3.8.

Hint.

Use Theorem 6.2.15, the Squeeze Theorem for sequences from Chapter 6. □

Returning to $\lim_{x\to 0} \frac{\sin x}{x}$ we'll see that the Squeeze Theorem is just what we need. First notice that since $D(x) = \sin x/x$ is an even function, we only need to focus on $x > 0$ in our inequalities. Consider the unit circle.

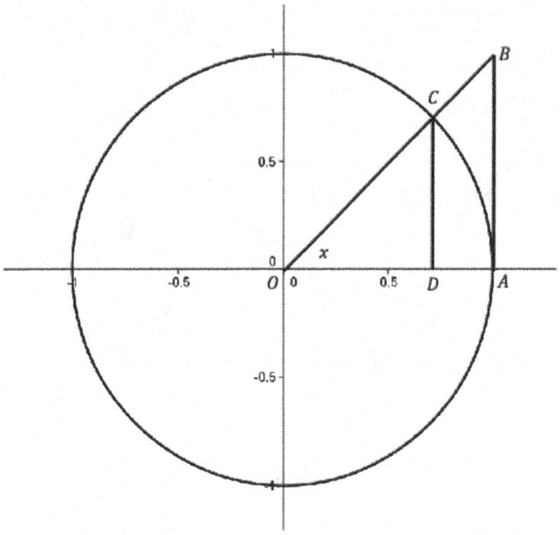

Problem 8.3.10 Use the fact that

$$\text{area } (\Delta OAC) < \text{ area (sector } OAC) < \text{ area } (\Delta OAB)$$

to show that if $0 < x < \pi/2$, then $\cos x < \sin x/x < 1$. Use the fact that all of these functions are even to extend the inequality for $-\pi/2 < x < 0$ and use the Squeeze Theorem to show $\lim_{x\to 0} \frac{\sin x}{x} = 1$. □

8.4 The Derivative, An Afterthought

No, the derivative isn't really an afterthought. Along with the integral it is, in fact, one of the most powerful and useful mathematical objects ever devised and we've been working very hard to provide a solid, rigorous foundation for it. In that sense it is a primary focus of our investigations.

On the other hand, now that we have built up all of the machinery we need to define and explore the concept of the derivative it will appear rather pedestrian alongside ideas like the convergence of power series, Fourier series, and the bizarre properties of $\mathbb{Q}$ and $\mathbb{R}$.

You spent an entire semester learning about the properties of the derivative and how to use them to explore the properties of functions so we will not repeat that effort here. Instead we will define it formally in terms of the ideas and techniques we've developed thus far.

Definition 8.4.1 The Derivative. Given a function $f(x)$ defined on an interval (a, b) we define

$$f'(x) = \lim_{h \to 0} \frac{f(x+h) - f(x)}{h}.$$

◊

There are a few fairly obvious facts about this definition which are nevertheless worth noticing explicitly:

1. The derivative is *defined at a point.* If it is defined at every point in an interval (a, b) then we say that the derivative exists at every point on the interval.

2. Since it is defined at a point it is at least theoretically possible for a function to be differentiable at a single point in its entire domain.

3. Since it is defined as a limit and not all limits exist, functions are not necessarily differentiable.

4. Since it is defined as a limit, Corollary 8.3.5 applies. That is, $f'(x)$ exists if and only if $\forall$ sequences (h_n), $h_n \neq 0$, if $\lim_{n\to\infty} h_n = 0$ then

   $$f'(x) = \lim_{n \to \infty} \frac{f(x+h_n) - f(x)}{h_n}.$$

   Since $\lim_{n\to\infty} h_n = 0$ this could also be written as

   $$f'(x) = \lim_{h_n \to 0} \frac{f(x+h_n) - f(x)}{h_n}.$$

Theorem 8.4.2 ***Differentiability Implies Continuity***
If f is differentiable at a point c then f is continuous at c as well.

Problem 8.4.3 Prove Theorem 8.4.2 □

As we mentioned, the derivative is an extraordinarily useful mathematical tool but it is not our intention to learn to *use* it here. Our purpose here is to define it rigorously (done) and to show that our formal definition does in fact recover the useful properties you came to know and love in your calculus course.

The first such property is known as Fermat's Theorem.

Theorem 8.4.4 Fermat's Theorem. *Suppose f is differentiable in some interval (a, b) containing c. If $f(c) \geq f(x)$ for every x in (a, b), then $f'(c) = 0$.*

Proof. Since $f'(c)$ exists we know that if $(h_n)_{n=1}^{\infty}$ converges to zero then the sequence $a_n = \frac{f(c+h_n)-f(c)}{h_n}$ converges to $f'(c)$. The proof consists of showing that $f'(c) \leq 0$ *and* that $f'(c) \geq 0$ from which we conclude that $f'(c) = 0$. We will only show the first part. The second is left as an exercise.

Claim: $f'(c) \leq 0$.

Let n_0 be sufficiently large that $\frac{1}{n_0} < b - c$ and take $(h_n) = \left(\frac{1}{n}\right)_{n=n_0}^{\infty}$. Then

$f\left(c+\frac{1}{n}\right)-f(c)\leq 0$ and $\frac{1}{n}>0$, so that

$$\frac{f\left(c+h_n\right)-f(c)}{h_n}\leq 0, \quad \forall n=n_0,n_0+1,\ldots$$

Therefore $f'(c)=\lim_{h_n\to 0}\frac{f\left(c+h_n\right)-f(c)}{h_n}\leq 0$ also. ■

Problem 8.4.5 Show that $f'(c)\geq 0$ and conclude that $f'(c)=0$. □

Problem 8.4.6 Show that if $f(c)\leq f(x)$ for all x in some interval (a,b) then $f'(c)=0$ too. □

Many of the most important properties of the derivative follow from what is called the Mean Value Theorem (**MVT**) which we now state.

Theorem 8.4.7 ***The Mean Value Theorem***

Suppose f' exists for every $x\in(a,b)$ and f is continuous on $[a,b]$. Then there is a real number $c\in(a,b)$ such that

$$f'(c)=\frac{f(b)-f(a)}{b-a}.$$

However, it would be difficult to prove the MVT right now. So we will first state and prove Rolle's Theorem, which can be seen as a special case of the MVT. The proof of the MVT will then follow easily.

Michel Rolle first stated the following theorem in 1691. Given this date and the nature of the theorem it would be reasonable to suppose that Rolle was one of the early developers of calculus but this is not so. In fact, Rolle was disdainful of both Newton and Leibniz's versions of calculus, once deriding them as a collection of "ingenious fallacies." It is a bit ironic that his theorem is so fundamental to the modern development of the calculus he ridiculed.

Theorem 8.4.8 ***Rolle's Theorem***

Suppose f' exists for every $x\in(a,b)$, f is continuous on $[a,b]$, and

$$f(a)=f(b).$$

Then there is a real number $c\in(a,b)$ such that

$$f'(c)=0.$$

Proof.

Comment. Any proof that relies on the Extreme Value Theorem is not complete until the EVT has been proved. We'll get to this in Chapter 9.

Since f is continuous on $[a,b]$ we see, by the Extreme Value Theorem, that f has both a maximum and a minimum on $[a,b]$. Denote the maximum by M and the minimum by m. There are several cases:

Case 1: $f(a)=f(b)=M=m$. In this case $f(x)$ is constant (why?). Therefore $f'(x)=0$ for every $x\in(a,b)$.

Case 2: $f(a)=f(b)=M\neq m$. In this case there is a real number $c\in(a,b)$ such that $f(c)$ is a local minimum. By Fermat's Theorem, $f'(c)=0$.

Case 3: $f(a)=f(b)=m\neq M$. In this case there is a real number $c\in(a,b)$ such that $f(c)$ is a local maximum. By Fermat's Theorem, $f'(c)=0$.

Case 4: $f(a) = f(b)$ is neither a maximum nor a minimum. In this case there is a real number $c_1 \in (a,b)$ such that $f(c_1)$ is a local maximum, and a real number $c_2 \in (a,b)$ such that $f(c_2)$ is a local minimum. By Fermat's Theorem, $f'(c_1) = f'(c_2) = 0$.

■

With Rolle's Theorem in hand we can prove the MVT which is really a corollary to Rolle's Theorem or, more precisely, it is a generalization of Rolle's Theorem. To prove it we only need to find the right function to apply Rolle's Theorem to. The following figure shows a function, $f(x)$, cut by a secant line, $L(x)$, from $(a, f(a))$ to $(b, f(b))$.

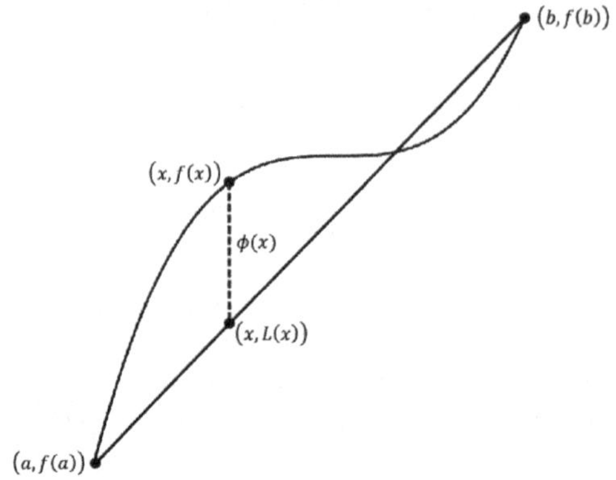

The vertical difference from $f(x)$ to the secant line, indicated by $\phi(x)$ in the figure should do the trick. You take it from there.

Problem 8.4.9 Prove the Mean Value Theorem. □

The Mean Value Theorem is extraordinarily useful. Almost all of the properties of the derivative that you used in calculus follow more or less easily from it. For example the following is true.

Corollary 8.4.10 *If $f'(x) > 0$ for every x in the interval (a,b) then for every $c, d \in (a,b)$ where $d > c$ we have*

$$f(d) > f(c).$$

That is, f is increasing on (a,b).

Proof. Suppose c and d are as described in the corollary. Then by the Mean Value Theorem there is some number, say $\alpha \in (c,d) \subseteq (a,b)$ such that

$$f'(\alpha) = \frac{f(d) - f(c)}{d - c}.$$

Since $f'(\alpha) > 0$ and $d - c > 0$ we have $f(d) - f(c) > 0$, or $f(d) > f(c)$. ■

Problem 8.4.11 Show that if $f'(x) < 0$ for every x in the interval (a,b) then f is decreasing on (a,b). □

Corollary 8.4.12 *Suppose f is differentiable on some interval (a,b), f' is continuous on (a,b), and that $f'(c) > 0$ for some $c \in (a,b)$. Then there is an interval, $I \subset (a,b)$, containing c such that for every x, y in I where $x \geq y$, $f(x) \geq f(y)$.*

Problem 8.4.13 Prove Corollary 8.4.12. □

Problem 8.4.14 Show that if f is differentiable on some interval (a, b) and that $f'(c) < 0$ for some $c \in (a, b)$ then there is an interval, $I \subset (a, b)$, containing c such that for every x, y in I where $x \geq y$, $f(x) \leq f(y)$. □

8.5 Additional Problems

Problem 8.5.1 Use the definition of continuity to prove that the constant function $g(x) = c$ is continuous at any point a. □

Problem 8.5.2

(a) Use the definition of continuity to prove that $\ln x$ is continuous at 1.

Hint.

You may want to use the fact $|\ln x| < \varepsilon \Leftrightarrow -\varepsilon < \ln x < \varepsilon$ to find a δ.

(b) Use part (a) to prove that $\ln x$ is continuous at any positive real number a.

Hint.

$\ln(x) = \ln(x/a) + \ln(a)$. This is a combination of functions which are continuous at a. Be sure to explain how you know that $\ln(x/a)$ is continuous at a.

□

Problem 8.5.3 Write a formal definition of the statement f is not continuous at a, and use it to prove that the function $f(x) = \begin{cases} x & \text{if } x \neq 1 \\ 0 & \text{if } x = 1 \end{cases}$ is not continuous at $a = 1$. □

Chapter 9

Intermediate and Extreme Values

9.1 Completeness of the Real Number System

Recall that in deriving the Lagrange and Cauchy forms of the remainder for Taylor series, we made use of the Extreme Value Theorem (**EVT**) and Intermediate Value Theorem (**IVT**). In Chapter 8, we produced an analytic definition of continuity that we can use to prove these theorems. To provide the rest of the necessary tools we need to explore the make-up of the real number system. To illustrate what we mean, suppose that we only used the rational number system. We could still use our definition of continuity and could still consider continuous functions such as $f(x) = x^2$. Notice that 2 is a value that lies between $f(1) = 1$ and $f(2) = 4$.

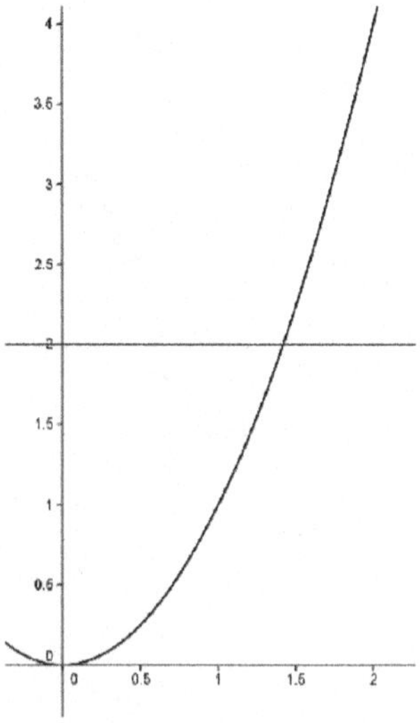

The IVT says that somewhere between 1 and 2, f must take on the value 2. That is, there must exist some number $c \in [1, 2]$ such that $f(c) = 2$. You might say, "Big deal! Everyone knows $c = \sqrt{2}$ works."

However, we are only working with rational numbers and $\sqrt{2}$ is not rational. As we saw in Chapter 2 the rational number system has holes in it, whereas the real number system doesn't. Again, "Big deal! Let's just say that the real number system contains (square) roots."

This sounds reasonable and it actually works for square roots, but consider the function $f(x) = x - \cos x$. We know this is a continuous function. We also know that $f(0) = -1$ and $f(\frac{\pi}{2}) = \frac{\pi}{2}$. According to the IVT, there should be some number $c \in [0, \frac{\pi}{2}]$, where $f(c) = 0$. The graph is below.

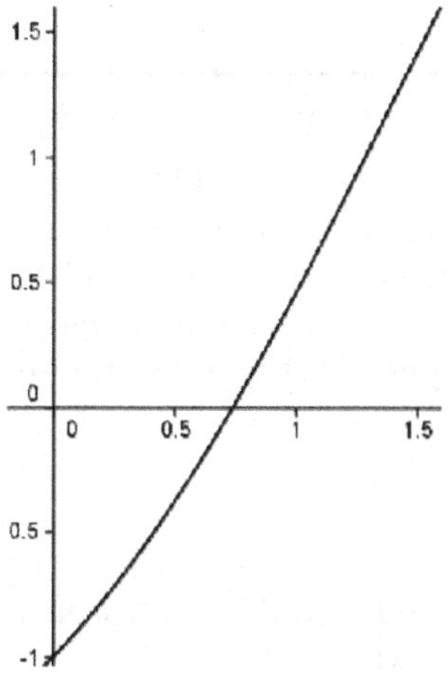

Graph of $f(x) = x - \cos x$

The situation is not as transparent as before. What would this mysterious c be where the curve crosses the x axis? Somehow we need to convey the idea that the real number system is a continuum. That is, it has no "holes" in it.

How about this? Why don't we just say that it has no holes in it? Sometimes the simple answer works best! But not in this case. How are we going to formulate a rigorous proof based on this statement? Just like with convergence and continuity, what we need is a rigorous way to convey this idea that the real number system does not have any holes, that it is *complete.*

We will see that there are several different, but equivalent ways to convey this notion of completeness. We will explore some of them in this chapter. For now we adopt the following as our *Completeness Axiom* for the real number system.

Axiom 9.1.1 Nested Interval Property of the Real Number System (NIP). *Suppose we have two sequences of real numbers $(x_n)$and (y_n) satisfying the following conditions:*

1. $x_1 \le x_2 \le x_3 \le \ldots$ *(this says that the sequence, (x_n), is non-decreasing)*
2. $y_1 \ge y_2 \ge y_3 \ge \ldots$ *(this says that the sequence, (y_n), is non-increasing)*
3. $\forall\ n,\ x_n \le y_n$
4. $\lim_{n\to\infty} (y_n - x_n) = 0$

Then there exists a unique number c such that $x_n \le c \le y_n$ for all n.

Geometrically, we have the following situation.

Notice that we have two sequences (x_n) and (y_n), one increasing (really non-decreasing) and one decreasing (non-increasing). These sequences do not pass each other. In fact, the following is true:

Problem 9.1.2 Let $(x_n), (y_n)$ be sequences as in the NIP. Show that for all $n, m \in \mathbb{N}$, $x_n \leq y_m$. □

They are also coming together in the sense that $\lim_{n\to\infty}(y_n - x_n) = 0$. The NIP says that in this case there is a unique real number c in the middle of all of this $x_n \leq c \leq y_n$ for all $n \in \mathbb{N}$.

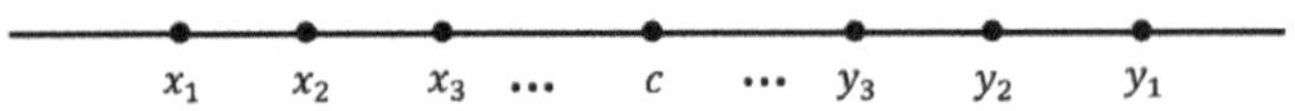

If there was no such c then there would be a hole where these two sequences come together. The NIP guarantees that there is no such hole. We do not need to prove this since an axiom is, by definition, a self evident truth. We are taking it on faith that the real number system obeys this law. The next problem shows that the completeness property distinguishes the real number system from the rational number system.

Problem 9.1.3

(a) Find two sequences of rational numbers $(x_n)$and (y_n) which satisfy properties 1-4 of the NIP and such that there is no rational number c satisfying the conclusion of the NIP.

Hint.

Consider the decimal expansion of an irrational number.

(b) Find two sequences of rational numbers (x_n) and (y_n) which satisfy properties 1-4 of the NIP and such that there is a rational number c satisfying the conclusion of the NIP.

□

You might find the name "Nested Interval Property" to be somewhat curious. One way to think about this property is to consider that we have a sequence of "nested closed intervals" $[x_1, y_1] \supseteq [x_2, y_2] \supseteq [x_3, y_3] \supseteq \cdots$ whose lengths $y_n - x_n$ are "shrinking to 0." The conclusion is that the intersection of these intervals is non-empty and, in fact, consists of a single point. That is, $\bigcap_{n=1}^{\infty}[x_n, y_n] = \{c\}$.

It appears that the sequences (x_n) and (y_n) in the NIP converge to c. This is, in fact, true and can be proven rigorously. In what follows, this will prove to be a valuable piece of information.

Theorem 9.1.4 *Suppose that we have two sequences (x_n) and (y_n) satisfying all of the assumptions of the Nested Interval Property. If c is the unique number such that $x_n \leq c \leq y_n$ for all n, then $\lim_{n\to\infty} x_n = c$ and $\lim_{n\to\infty} y_n = c$.*

Problem 9.1.5 Prove Theorem 9.1.4. □

To illustrate the idea that the NIP "plugs the holes" in the real line, we will prove the existence of square roots of nonnegative real numbers.

Theorem 9.1.6 *Suppose $a \in \mathbb{R}$, $a \geq 0$. There exists a real number $c \geq 0$ such that $c^2 = a$.*

Notice that we can't just say, "Let $c = \sqrt{a}$," since the idea is to show that this square root exists. In fact, throughout this proof, we cannot really use a square root symbol as we haven't yet proved that they (square roots) exist. We will give the idea behind the proof as it illustrates how the NIP is used.

Sketch of Proof. Our strategy is to construct two sequences which will "narrow in" on the number c that we seek. With that in mind, we need to find a number x_1 such that $x_1^2 \leq a$ and a number y_1 such that $y_1^2 \geq a$. (Remember that we can't say $\sqrt{x_1}$ or$\sqrt{y_1}$.) There are many possibilities, but how about $x_1 = 0$ and

$y_1 = a + 1$? You can check that these will satisfy $x_1^2 \le a \le y_1^2$. Furthermore $x_1 \le y_1$. This is the starting point.
The technique we will employ is often called a bisection technique, and is a useful way to set ourselves up for applying the NIP. Let m_1 be the midpoint of the interval $[x_1, y_1]$. Then either we have $m_1^2 \le a$ or $m_1^2 \ge a$. In the case $m_1^2 \le a$, we really want m_1 to take the place of x_1 since it is larger than x_1, but still represents an underestimate for what would be the square root of a. This thinking prompts the following move. If $m_1^2 \le a$, we will relabel things by letting $x_2 = m_1$ and $y_2 = y_1$. The situation looks like this on the number line.

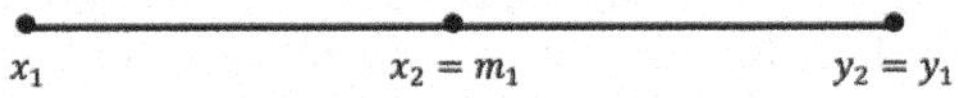

In the other case where $a \le m_1^2$, we will relabel things by letting $x_2 = x_1$ and $y_2 = m_1$. The situation looks like this on the number line.

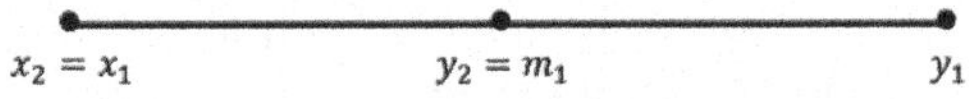

In either case, we've reduced the length of the interval where the square root lies to half the size it was before. Stated in more specific terms, in either case we have the same results:

$$x_1 \le x_2 \le y_2 \le y_1; \quad x_1^2 \le a \le y_1^2; \quad x_2^2 \le a \le y_2^2$$

and

$$y_2 - x_2 = \frac{1}{2}(y_1 - x_1).$$

Now we play the same game, but instead we start with the interval $[x_2, y_2]$. Let m_2 be the midpoint of $[x_2, y_2]$. Then we have $m_2^2 \le a$ or $m_2^2 \ge a$. If $m_2^2 \le a$, we relabel $x_3 = m_2$ and $y_3 = y_2$. If $a \le m_2^2$, we relabel $x_3 = x_2$ and $y_3 = m_2$. In either case, we end up with

$$x_1 \le x_2 \le x_3 \le y_3 \le y_2 \le y_1; \quad x_1^2 \le a \le y_1^2; \quad x_2^2 \le a \le y_2^2; \quad x_3^2 \le a \le y_3^2$$

and

$$y_3 - x_3 = \frac{1}{2}(y_2 - x_2) = \frac{1}{2^2}(y_1 - x_1).$$

Continuing in this manner, we will produce two sequences, $(x_n)$and (y_n) satisfying the following conditions:

1. $x_1 \le x_2 \le x_3 \le \ldots$
2. $y_1 \ge y_2 \ge y_3 \ge \ldots$
3. $\forall\ n,\ x_n \le y_n$
4. $\lim_{n\to\infty}(y_n - x_n) = \lim_{n\to\infty}\frac{1}{2^{n-1}}(y_1 - x_1) = 0$
5. These sequences also satisfy the following property:

$$\forall n,\ x_n^2 \le a \le y_n^2$$

Properties 1-4 tell us that $(x_n)$and (y_n) satisfy all of the conditions of the NIP, so we can conclude that there must exist a real number c such that $x_n \le c \le y_n$ for all n. At this point, you should be able to use property 5. to show that $c^2 = a$ as desired. ■

Problem 9.1.7 Turn the above outline into a formal proof of Theorem 9.1.6. □

The bisection method we employed in the proof of Theorem 9.1.6 is pretty typical of how we will use the NIP, as taking midpoints ensures that we will create a sequence of "nested intervals." We will employ this strategy in the proofs of the IVT and EVT. Deciding how to relabel the endpoints of our intervals will be determined by what we want to do with these two sequences of real numbers. This will typically lead to a fifth property, which will be crucial in proving that the c guaranteed by the NIP does what we want it to do. Specifically, in the above example, we always wanted our candidate for $\sqrt{a}$ to be in the interval $[x_n, y_n]$. This judicious choice led to the extra Property 5: $\forall\ n,\ x_n^2 \leq a \leq y_n^2$. In applying the NIP to prove the IVT and EVT, we will find that properties 1-4 will stay the same. Property 5 is what will change based on the property we want c to have.

Before we tackle the IVT and EVT, let's use the NIP to address an interesting question about the Harmonic Series. Recall that the Harmonic Series, $1 + \frac{1}{2} + \frac{1}{3} + \frac{1}{4} + \cdots$, grows without bound, that is, $\sum_{n=1}^{\infty} \frac{1}{n} = \infty$. The question is how slowly does this series grow? For example, how many terms would it take before the series surpasses 100? 1000? 10000? Leonhard Euler decided to tackle this problem in the following way. Euler decided to consider the $\lim_{n\to\infty} \left(\left(1 + \frac{1}{2} + \frac{1}{3} + \cdots + \frac{1}{n}\right) - \ln\ (n+1)\right)$. This limit is called Euler's constant and is denoted by γ. This says that for n large, we have $1 + \frac{1}{2} + \frac{1}{3} + \cdots + \frac{1}{n} \approx \ln(n+1) + \gamma$. If we could approximate γ, then we could replace the inequality $1 + \frac{1}{2} + \frac{1}{3} + \cdots + \frac{1}{n} \geq 100$ with the more tractable inequality $\ln(n+1) + \gamma \geq 0$ and solve for n in this. This should tell us roughly how many terms would need to be added in the Harmonic Series to surpass 100. Approximating γ with a computer is not too bad. We could make n as large as we wish in $\left(1 + \frac{1}{2} + \frac{1}{3} + \cdots + \frac{1}{n}\right) - \ln(1+n)$ to make closer approximations for γ. The real issue is, ***HOW DO WE KNOW THAT***

$$\lim_{n\to\infty}\left(\left(1+\frac{1}{2}+\frac{1}{3}+\cdots+\frac{1}{n}\right) - \ln\ (\mathrm{n}+1)\right)$$

ACTUALLY EXISTS?

You might want to say that obviously it should, but let us point out that as of the printing of this book (2013), it is not even known if γ is rational or irrational. So, in our opinion the existence of this limit is not so obvious. This is where the NIP will come into play; we will use it to show that this limit, in fact, exists. The details are in the following problem.

Problem 9.1.8 The purpose of this problem is to show that

$$\lim_{n\to\infty}\left(\left(1+\frac{1}{2}+\frac{1}{3}+\cdots+\frac{1}{n}\right) - \ln(n+1)\right)$$

exists.

(a) Let $x_n = \left(1 + \frac{1}{2} + \frac{1}{3} + \cdots + \frac{1}{n}\right) - \ln(n+1)$. Use the following diagram to show

$$x_1 \leq x_2 \leq x_3 \leq \cdots$$

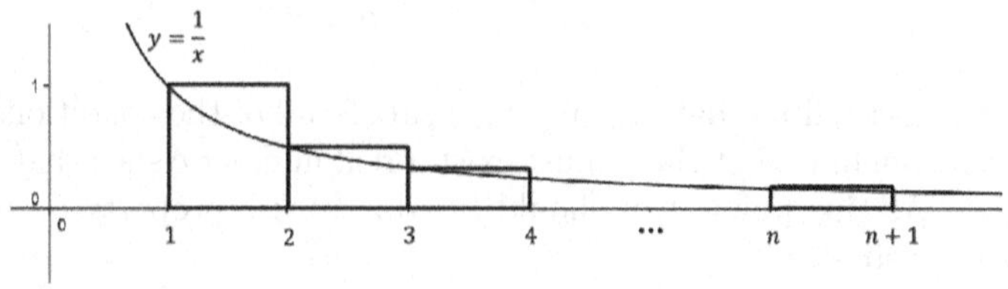

(b) Let $z_n = \ln(n+1) - \left(\frac{1}{2} + \frac{1}{3} + \cdots + \frac{1}{n+1}\right)$. Use a similar diagram to show that $z_1 \le z_2 \le z_3 \le \cdots$.

(c) Let $y_n = 1 - z_n$. Show that (x_n) and (y_n) satisfy the hypotheses of the nested interval property and use the NIP to conclude that there is a real number γ such that $x_n \le \gamma \le y_n$ for all n.

(d) Conclude that $\lim_{n\to\infty} \left(\left(1 + \frac{1}{2} + \frac{1}{3} + \cdots + \frac{1}{n}\right) - \ln(n+1)\right) = \gamma$.

□

Problem 9.1.9 Use the fact that $x_n \le \gamma \le y_n$ for all n to approximate γ to three decimal places. □

Problem 9.1.10

(a) Use the fact that for large n, $1 + \frac{1}{2} + \frac{1}{3} + \cdots + \frac{1}{n} \approx \ln(n+1) + \gamma$ to determine approximately how large n must be to make

$$1 + \frac{1}{2} + \frac{1}{3} + \cdots + \frac{1}{n} \ge 100.$$

(b) Suppose we have a supercomputer which can add 10 trillion terms of the Harmonic Series per second. Approximately how many Earth Ages (the current age of the Earth) would it take for this computer to sum the Harmonic Series until it surpasses 100?

□

9.2 Proof of the Intermediate Value Theorem

We now have all of the tools we need to prove the Intermediate Value Theorem (**IVT**).

Theorem 9.2.1 ***Intermediate Value Theorem (IVT)***
Suppose $f(x)$ is continuous on $[a, b]$ and v is any real number between $f(a)$ and $f(b)$. Then there exists a real number $c \in [a, b]$ such that $f(c) = v$.

Sketch of Proof. We have two cases to consider: $f(a) \le v \le f(b)$ and $f(a) \ge v \ge f(b)$.
We will look at the case $f(a) \le v \le f(b)$. Let $x_1 = a$ and $y_1 = b$, so we have $x_1 \le y_1$ and $f(x_1) \le v \le f(y_1)$. Let m_1 be the midpoint of $[x_1, y_1]$ and notice that we have either $f(m_1) \le v$ or $f(m_1) \ge v$. If $f(m_1) \le v$, then we relabel $x_2 = m_1$ and $y_2 = y_1$. If $f(m_1) \ge v$, then we relabel $x_2 = x_1$ and $y_2 = m_1$. In either case, we end up with $x_1 \le x_2 \le y_2 \le y_1$, $y_2 - x_2 = \frac{1}{2}(y_1 - x_1)$, $f(x_1) \le v \le f(y_1)$, and $f(x_2) \le v \le f(y_2)$.
Now play the same game with the interval $[x_2, y_2]$. If we keep playing this game, we will generate two sequences (x_n) and (y_n), satisfying all of the conditions of the nested interval property. These sequences will also satisfy the following extra property: $\forall\ n$, $f(x_n) \le v \le f(y_n)$. By the NIP, there exists a c such that $x_n \le c \le y_n$, $\forall\ n$. This should be the c that we seek though this is not obvious. Specifically, we need to show that $f(c) = v$. This should be where the continuity of f at c and the extra property on $(x_n)$and (y_n) come into play. ■

Problem 9.2.2 Turn the ideas of the previous paragraphs into a formal proof of the IVT for the case $f(a) \le v \le f(b)$. □

Problem 9.2.3 We can modify the proof of the case $f(a) \leq v \leq f(b)$ into a proof of the IVT for the case $f(a) \geq v \geq f(b)$. However, there is a sneakier way to prove this case by applying the IVT to the function $-f$. Do this to prove the IVT for the case $f(a) \geq v \geq f(b)$. □

Problem 9.2.4 Use the IVT to prove that any polynomial of odd degree must have a real root. □

9.3 The Bolzano-Weierstrass Theorem

Once we introduced the Nested Interval Property (Axiom 9.1.1), the Intermediate Value Theorem (Theorem 9.2.1) followed pretty readily. The proof of Extreme Value Theorem (Theorem 9.4.8) takes a bit more work. First we need to show that a function that satisfies the conditions of the EVT is bounded.

Theorem 9.3.1 *A continuous function defined on a closed, bounded interval must be bounded. That is, let f be a continuous function defined on $[a, b]$. Then there exists a positive real number B such that $|f(x)| \leq B$ for all $x \in [a, b]$.*

Sketch of Alleged Proof: Let's assume, for contradiction, that there is no such bound B. This says that for any positive integer n, there must exist $x_n \in [a, b]$ such that $|f(x_n)| > n$. (Otherwise n would be a bound for f.) ***IF*** the sequence (x_n) converged to something in $[a, b]$, say c, then we would have our contradiction. Indeed, we would have $\lim_{n\to\infty} x_n = c$. By the continuity of f at c and Theorem 8.2.1 of Chapter 8, we would have $\lim_{n\to\infty} f(x_n) = f(c)$. This would say that the sequence $(f(x_n))$ converges, so by Lemma 6.2.7 of Chapter 6, it must be bounded. This would provide our contradiction, as we had $|f(x_n)| > n$, for all positive integers n. ***QED?***

This would all work well except for one little problem. The way it was constructed, there is no reason to expect the sequence (x_n) to converge to anything and we can't make such an assumption. That is why we emphasized the ***IF*** above. Fortunately, this idea can be salvaged. While it is true that the sequence (x_n) may not converge, part of it will. We will need the following definition.

Definition 9.3.2 Let $(n_k)_{k=1}^{\infty}$ be a strictly increasing sequence of positive integers; that is, $n_1 < n_2 < n_3 < \cdots$. If $(x_n)_{n=1}^{\infty}$ is a sequence, then $(x_{n_k})_{k=1}^{\infty} = (x_{n_1}, x_{n_2}, x_{n_3}, \ldots)$ is called a subsequence of (x_n). ♢

The idea is that a subsequence of a sequence is a part of the sequence, (x_n), which is itself a sequence. However, it is a little more restrictive. We can choose any term in our sequence to be part of the subsequence, but once we choose that term, we can't go backwards. This is where the condition $n_1 < n_2 < n_3 < \cdots$ comes in. For example, suppose we started our subsequence with the term x_{100}. We could not choose our next term to be x_{99}. The subscript of the next term would have to be greater than 100. In fact, the thing about a subsequence is that it is all in the subscripts; we are really choosing a subsequence (n_k) of the sequence of subscripts (n) in (x_n).

Example 9.3.3 Given the sequence (x_n), the following are subsequences.

(a) $(x_2, x_4, x_6, \ldots) = (x_{2k})_{k=1}^{\infty}$

(b) $(x_1, x_4, x_9, \ldots) = (x_{k^2})_{k=1}^{\infty}$

(c) (x_n) itself.

□

Example 9.3.4 The following are ***NOT*** subsequences.

(a) $(x_1, x_1, x_1, \ldots)$

(b) $(x_{99}, x_{100}, x_{99}, \ldots)$

(c) (x_1, x_2, x_3)

□

The subscripts in the examples we have seen so far have a discernable pattern, but this need not be the case. For example,

$$(x_2, x_5, x_{12}, x_{14}, x_{23}, \ldots)$$

would be a subsequence as long as the subscripts form an increasing sequence themselves.

Problem 9.3.5 Suppose $\lim_{n\to\infty} x_n = c$. Prove that $\lim_{k\to\infty} x_{n_k} = c$ for any subsequence (x_{n_k}) of (x_n).

Hint.

First prove that $n_k \geq k$. □

A very important theorem about subsequences was introduced by Bernhard Bolzano and, later, independently proven by Karl Weierstrass. Basically, this theorem says that any bounded sequence of real numbers has a convergent subsequence.

Theorem 9.3.6 ***The Bolzano-Weierstrass Theorem***

Let (x_n) be a sequence of real numbers such that $x_n \in [a, b]$, $\forall$ n. Then there exists $c \in [a, b]$ and a subsequence (x_{n_k}), such that $\lim_{k\to\infty} x_{n_k} = c$.

As an example of this theorem, consider the sequence

$$((-1)^n) = (-1, 1, -1, 1, \ldots).$$

This sequence does not converge, but the subsequence

$$\left((-1)^{2k}\right) = (1, 1, 1, \ldots)$$

converges to 1. This is not the only convergent subsequence, as $\left((-1)^{2k+1}\right) = (-1, -1, -1, \ldots)$ converges to -1. Notice that if the sequence is unbounded, then all bets are off; the sequence may have a convergent subsequence or it may not. The sequences $(((-1)^n + 1)\, n)$ and (n) represent these possibilities as the first has, for example, $\left(\left((-1)^{2k+1} + 1\right)(2k + 1)\right) = (0, 0, 0, \ldots)$ and the second one has none.

The Bolzano-Weierstrass Theorem says that no matter how "random" the sequence (x_n) may be, as long as it is bounded then some part of it must converge. This is very useful when one has some process which produces a "random" sequence such as what we had in the idea of the alleged proof in Theorem 9.3.1.

Sketch of a Proof of the Bolzano-Weierstrass Theorem. Suppose we have our sequence (x_n) such that $x_n \in [a, b]$, $\forall$ n. To find our c for the subsequence to converge to we will use the NIP. Since we are already using (x_n) as our original sequence, we will need to use different letters in setting ourselves up for the NIP. With this in mind, let $a_1 = a$ and $b_1 = b$, and notice that $x_n \in [a_1, b_1]$ for infinitely many n. (This is, in fact true for all n, but you'll see why we said it the way we did.) Let m_1 be the midpoint of $[a_1, b_1]$ and notice that either $x_n \in [a_1, m_1]$ for infinitely many n or $x_n \in [m_1, b_1]$ for infinitely many n. If $x_n \in [a_1, m_1]$ for infinitely many n, then we relabel $a_2 = a_1$ and $b_2 = m_1$. If $x_n \in [m_1, b_1]$ for infinitely many n, then relabel $a_2 = m_1$ and $b_2 = b_1$. In either

case, we get $a_1 \le a_2 \le b_2 \le b_1$, $b_2 - a_2 = \frac{1}{2}(b_1 - a_1)$, and $x_n \in [a_2, b_2]$ for infinitely many n.
Now we consider the interval $[a_2, b_2]$ and let m_2 be the midpoint of $[a_2, b_2]$. Since $x_n \in [a_2, b_2]$ for infinitely many n, then either $x_n \in [a_2, m_2]$ for infinitely many n or $x_n \in [m_2, b_2]$ for infinitely many n. If $x_n \in [a_2, m_2]$ for infinitely many n, then we relabel $a_3 = a_2$ and $b_3 = m_2$. If $x_n \in [m_2, b_2]$ for infinitely many n, then we relabel $a_3 = m_2$ and $b_3 = b_2$. In either case, we get $a_1 \le a_2 \le a_3 \le b_3 \le b_2 \le b_1$, $b_3 - a_3 = \frac{1}{2}(b_2 - a_2) = \frac{1}{2^2}(b_1 - a_1)$, and $x_n \in [a_3, b_3]$ for infinitely many n.
If we continue in this manner, we will produce two sequences $(a_k)$and (b_k) with the following properties:

1. $a_1 \le a_2 \le a_3 \le \cdots$
2. $b_1 \ge b_2 \ge b_3 \ge \cdots$
3. $\forall\ k,\ a_k \le b_k$
4. $\lim\limits_{k\to\infty}(b_k - a_k) = \lim\limits_{k\to\infty}\dfrac{1}{2^{k-1}}(b_1 - a_1) = 0$
5. For each k, $x_n \in [a_k, b_k]$ for infinitely many n

By properties 1-5 and the NIP, there exists a unique c such that $c \in [a_k, b_k]$, for all k. In particular, $c \in [a_1, b_1] = [a, b]$.
We have our c. Now we need to construct a subsequence converging to it. Since $x_n \in [a_1, b_1]$ for infinitely many n, choose an integer n_1 such that $x_{n_1} \in [a_1, b_1]$. Since $x_n \in [a_2, b_2]$ for infinitely many n, choose an integer $n_2 > n_1$ such that $x_{n_2} \in [a_2, b_2]$. (Notice that to make a subsequence it is crucial that $n_2 > n_1$, and this is why we needed to insist that $x_n \in [a_2, b_2]$ for infinitely many n.) Continuing in this manner, we should be able to build a subsequence (x_{n_k}) that will converge to c. You can supply the details in the following problem. ■

Problem 9.3.7 Turn the ideas of the above outline into a formal proof of the Bolzano-Weierstrass Theorem. □

Problem 9.3.8 Use the Bolzano-Weierstrass Theorem to complete the proof of Theorem 9.3.1. □

9.4 The Supremum and the Extreme Value Theorem

Theorem 9.3.1 says that a continuous function on a closed, bounded interval must be bounded. Boundedness, in and of itself, does not ensure the existence of a maximum or minimum. We must also have a closed, bounded interval. To illustrate this, consider the continuous function $f(x) = \tan^{-1}x$ defined on the (unbounded) interval $(-\infty, \infty)$.

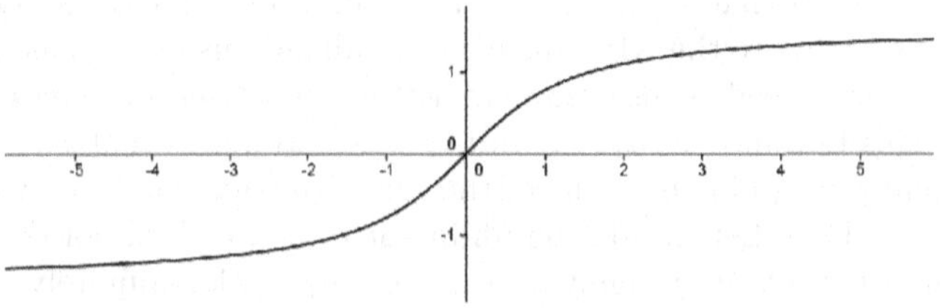

$f(x) = \tan^{-1} x$

This function is bounded between $-\frac{\pi}{2}$and $\frac{\pi}{2}$, but it does not attain a maximum or minimum as the lines $y = \pm\frac{\pi}{2}$ are horizontal asymptotes. Notice that if we restricted the domain to a closed, bounded interval then it would attain its extreme values on that interval (as guaranteed by the **EVT**).

To find a maximum we need to find the smallest possible upper bound for the range of the function. This prompts the following definitions.

Definition 9.4.1 Let $S \subseteq \mathbb{R}$ and let b be a real number. We say that b is an upper bound of S provided $b \geq x$ for all $x \in S$. ♢

For example, if $S = (0, 1)$, then any b with $b \geq 1$ would be an upper bound of S. Furthermore, the fact that b is not an element of the set S is immaterial. Indeed, if $T = [0, 1]$, then any b with $b \geq 1$ would still be an upper bound of T. Notice that, in general, if a set has an upper bound, then it has infinitely many since any number larger than that upper bound would also be an upper bound. However, there is something special about the smallest upper bound.

Definition 9.4.2 Least Upper Bound Property (LUBP). Let $S \subseteq \mathbb{R}$ and let b be a real number. We say that b is the *least upper bound* of S provided

a $b \geq x$ for all $x \in S$. (b is an upper bound of S)

b If $c \geq x$ for all $x \in S$, then $c \geq b$. (Any upper bound of S is at least as big as b.)

In this case, we also say that b *is the supremum* of S and we write

$$b = \sup(S).$$

♢

Notice that the definition really says that b is the smallest upper bound of S. Also notice that the second condition can be replaced by its contrapositive so we can say that $b = \sup S$ if and only if

a $b \geq x$ for all $x \in S$

b If $c < b$ then there exists $x \in S$ such that $c < x$.

The second condition says that if a number c is less than b, then it can't be an upper bound, so that b really is the smallest upper bound.

Also notice that the supremum of the set may or may not be in the set itself. This is illustrated by the examples above as in both cases, $1 = \sup(0, 1)$ and $1 = \sup[0, 1]$. Obviously, a set which is not bounded above such as $\mathbb{N} = \{1, 2, 3, \ldots\}$ cannot have a supremum. However, for non-empty sets which are bounded above, we have the following.

Theorem 9.4.3 The Least Upper Bound Property (LUBP). *Let S be a non-empty subset of $\mathbb{R}$ which is bounded above. Then S has a supremum.*

Sketch of Proof. Since $S \neq \emptyset$, then there exists $s \in S$. Since S is bounded above then it has an upper bound, say b. We will set ourselves up to use the Nested Interval Property. With this in mind, let $x_1 = s$ and $y_1 = b$ and notice that $\exists\, x \in S$ such that $x \geq x_1$ (namely, x_1 itself) and $\forall x \in S$, $y_1 \geq x$. You probably guessed what's coming next: let m_1 be the midpoint of $[x_1, y_1]$. Notice that either $m_1 \geq x, \forall x \in S$ or $\exists\, x \in S$ such that $x \geq m_1$. In the former case, we relabel, letting $x_2 = x_1$ and $y_2 = m_1$. In the latter case, we let $x_2 = m_1$ and $y_2 = y_1$. In either case, we end up with $x_1 \leq x_2 \leq y_2 \leq y_1$, $y_2 - x_2 = \frac{1}{2}(y_1 - x_1)$, and $\exists\, x \in S$ such that $x \geq x_2$ and $\forall x \in S$, $y_2 \geq x$. If we continue this process, we end up with two sequences, $(x_n)$and (y_n), satisfying the following conditions:

1. $x_1 \le x_2 \le x_3 \le \ldots$

2. $y_1 \ge y_2 \ge y_3 \ge \ldots$

3. $\forall\, n,\ x_n \le y_n$

4. $\lim\limits_{n\to\infty} (y_n - x_n) = \lim\limits_{n\to\infty} \frac{1}{2^{n-1}} (y_1 - x_1) = 0$

5. $\forall\, n, \exists\, x \in S$ such that $x \ge x_n$ and $\forall x \in S,\ y_n \ge x$,

By properties 1-5 and the NIP there exists c such that $x_n \le c \le y_n$, $\forall\, n$. We will leave it to you to use property 5 to show that $c = \sup S$. ■

Problem 9.4.4 Complete the above ideas to provide a formal proof of Theorem 9.4.3. □

Notice that we really used the fact that S was non-empty and bounded above in the proof of Theorem 9.4.3. This makes sense, since a set which is not bounded above cannot possibly have a least upper bound. In fact, any real number is an upper bound of the empty set so that the empty set would not have a least upper bound.

The following corollary to Theorem 9.4.3 can be very useful.

Corollary 9.4.5 *Let (x_n) be a bounded, increasing sequence of real numbers. That is, $x_1 \le x_2 \le x_3 \le \cdots$. Then (x_n) converges to some real number c.*

Problem 9.4.6 Prove Corollary 9.4.5.

Hint.

Let $c = \sup\{x_n | \, n = 1, 2, 3, \ldots\}$. To show that $\lim\limits_{n\to\infty} x_n = c$, let $\epsilon > 0$.Note that $c - \epsilon$ is not an upper bound. You take it from here! □

Problem 9.4.7 Consider the following curious expression

$$\sqrt{2+\sqrt{2+\sqrt{2+\sqrt{\ldots}}}}$$

We will use Corollary 9.4.5 to show that this actually converges to some real number. After we know it converges we can actually compute what it is. Of course to do so, we need to define things a bit more precisely. With this in mind consider the following sequence (x_n) defined as follows:

$$x_1 = \sqrt{2}$$

$$x_{n+1} = \sqrt{2 + x_n}.$$

a Use induction to show that $x_n < 2$ for $n = 1, 2, 3, \ldots$.

b Use the result from part (a) to show that $x_n < x_{n+1}$ for $n = 1, 2, 3, \ldots$.

c From Corollary 9.4.5, we have that (x_n) must converge to some number c. Use the fact that (x_{n+1}) must converge to c as well to compute what c must be.

□

We now have all the tools we need to tackle the Extreme Value Theorem.

Theorem 9.4.8 ***Extreme Value Theorem (EVT)***

Suppose f is continuous on $[a, b]$. Then there exists $c, d \in [a, b]$ such that $f(d) \le f(x) \le f(c)$, for all $x \in [a, b]$.

Sketch of Proof. We will first show that f attains its maximum. To this end, recall that Theorem 9.3.1 tells us that $f[a,b] = \{f(x) | x \in [a,b]\}$ is a bounded set. By the LUBP, $f[a,b]$ must have a least upper bound which we will label s, so that $s = \sup f[a,b]$. This says that $s \geq f(x)$,for all $x \in [a,b]$. All we need to do now is find a $c \in [a,b]$ with $f(c) = s$. With this in mind, notice that since $s = \sup f[a,b]$, then for any positive integer n, $s - \frac{1}{n}$ is not an upper bound of $f[a,b]$. Thus there exists $x_n \in [a,b]$ with $s - \frac{1}{n} < f(x_n) \leq s$. Now, by the Bolzano-Weierstrass Theorem, (x_n) has a convergent subsequence (x_{n_k}) converging to some $c \in [a,b]$. Using the continuity of f at c, you should be able to show that $f(c) = s$. To find the minimum of f, find the maximum of $-f$. ■

Problem 9.4.9 Formalize the above ideas into a proof of Theorem 9.4.8. □

Notice that we used the NIP to prove both the Bolzano-Weierstrass Theorem and the LUBP. This is really unavoidable, as it turns out that all of those statements are equivalent in the sense that any one of them can be taken as the completeness axiom for the real number system and the others proved as theorems. This is not uncommon in mathematics, as people tend to gravitate toward ideas that suit the particular problem they are working on. In this case, people realized at some point that they needed some sort of completeness property for the real number system to prove various theorems. Each individual's formulation of completeness fit in with his understanding of the problem at hand. Only in hindsight do we see that they were really talking about the same concept: the completeness of the real number system. In point of fact, most modern textbooks use the LUBP as the axiom of completeness and prove all other formulations as theorems. We will finish this section by showing that either the Bolzano-Weierstrass Theorem or the LUBP can be used to prove the NIP. This says that they are all equivalent and that any one of them could be taken as the completeness axiom.

Problem 9.4.10 Use the Bolzano-Weierstrass Theorem to prove the NIP. That is, assume that the Bolzano-Weierstrass Theorem holds and suppose we have two sequences of real numbers, (x_n) and (y_n), satisfying:

1. $x_1 \leq x_2 \leq x_3 \leq \ldots$
2. $y_1 \geq y_2 \geq y_3 \geq \ldots$
3. $\forall\, n,\ x_n \leq y_n$
4. $\lim_{n\to\infty} (y_n - x_n) = 0.$

Prove that there is a real number c such that $x_n \leq c \leq y_n$, for all n. □

Since the Bolzano-Weierstrass Theorem and the Nested Interval Property are equivalent, it follows that the Bolzano-Weierstrass Theorem will not work for the rational number system.

Problem 9.4.11 Find a bounded sequence of rational numbers such that no subsequence of it converges to a rational number. □

Problem 9.4.12 Use the Least Upper Bound Property to prove the Nested Interval Property. That is, assume that every non-empty subset of the real numbers which is bounded above has a least upper bound; and suppose that we have two sequences of real numbers (x_n) and (y_n), satisfying:

1. $x_1 \leq x_2 \leq x_3 \leq \ldots$
2. $y_1 \geq y_2 \geq y_3 \geq \ldots$

3. $\forall\, n, x_n \leq y_n$

4. $\lim_{n\to\infty} (y_n - x_n) = 0.$

Prove that there exists a real number c such that $x_n \leq c \leq y_n$, for all n. (Again, the c will, of necessity, be unique, but don't worry about that.)

Hint.

Corollary 9.4.5 might work well here. □

Problem 9.4.13 Since the LUBP is equivalent to the NIP it does not hold for the rational number system. Demonstrate this by finding a non-empty set of rational numbers which is bounded above, but whose supremum is an irrational number. □

We have the machinery in place to clean up a matter that was introduced in Chapter 2. If you recall (or look back) we introduced the Archimedean Property of the real number system. This property says that given any two positive real numbers a, b, there exists a positive integer n with $na > b$. As we mentioned in Chapter 2, this was taken to be intuitively obvious. The analogy we used there was to emptying an ocean b with a teaspoon a provided we are willing to use it enough times n. The completeness of the real number system allows us to prove it as a formal theorem.

Theorem 9.4.14 ***Archimedean Property of*** $\mathbb{R}$

Given any positive real numbers a and b, there exists a positive integer n, such that $na > b$.

Problem 9.4.15 Prove Theorem 9.4.14.

Hint.

Assume that there are positive real numbers a and b, such that $na \leq b$ $\forall n \in \mathbb{N}$. Then $\mathbb{N}$ would be bounded above by b/a. Let $s = \sup(\mathbb{N})$ and consider $s - 1$. □

Given what we've been doing, one might ask if the Archimedean Property is equivalent to the LUBP (and thus could be taken as an axiom). The answer lies in the following problem.

Problem 9.4.16 Does $\mathbb{Q}$ satisfy the Archimedean Property and what does this have to do with the question of taking the Archimedean Property as an axiom of completeness? □

9.5 Additional Problems

Problem 9.5.1 Mimic the definitions of an upper bound of a set and the least upper bound (supremum) of a set to give definitions for a lower bound of a set and the greatest lower bound (infimum) of a set.

Note: The infimum of a set S is denoted by $\inf(S)$. □

Problem 9.5.2 Find the least upper bound (supremum) and greatest lower bound (infimum) of the following sets of real numbers, if they exist. (If one does not exist then say so.)

a $S = \left\{ \frac{1}{n} \,|\, n = 1, 2, 3, \ldots \right\}$

b $T = \{ r \,|\, r \text{ is rational and } r^2 < 2 \}$

c $(-\infty, 0) \cup (1, \infty)$

d $R = \left\{ \frac{(-1)^n}{n} \mid n = 1, 2, 3, \ldots \right\}$

e $(2, 3\pi] \cap \mathbb{Q}$

f The empty set $\emptyset$

$\square$

Problem 9.5.3 Let $S \subseteq \mathbb{R}$ and let $T = \{-x \mid x \in S\}$.

a Prove that b is an upper bound of S if and only if $-b$ is a lower bound of T.

b Prove that $b = \sup S$ if and only if $-b = \inf T$.

$\square$

Chapter 10

Back to Power Series

10.1 Uniform Convergence

We have developed precise analytic definitions of the convergence of a sequence and continuity of a function and we have used these to prove the EVT and IVT for a continuous function. We will now draw our attention back to the question that originally motivated these definitions, "Why are Taylor series well behaved, but Fourier series are not necessarily?" More precisely, we mentioned that whenever a power series converges then whatever it converged to was continuous. Moreover, if we differentiate or integrate these series term by term then the resulting series will converge to the derivative or integral of the original series. This was not always the case for Fourier series. For example consider the function

$$
\begin{aligned}
f(x) &= \frac{4}{\pi}\left(\sum_{k=0}^{\infty}\frac{(-1)^k}{2k+1}\cos\left((2k+1)\pi x\right)\right) \\
&= \frac{4}{\pi}\left(\cos(\pi x) - \frac{1}{3}\cos(3\pi x) + \frac{1}{5}(5\pi x) - \ldots\right).
\end{aligned}
$$

We have seen that the graph of f is given by

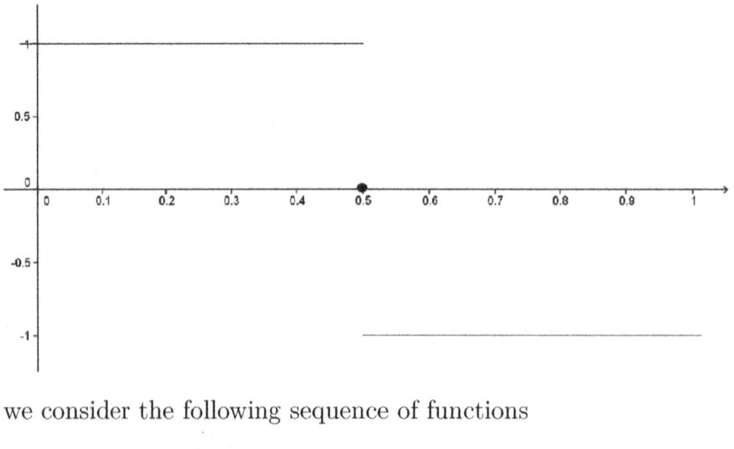

If we consider the following sequence of functions

$$
\begin{aligned}
f_1(x) =& \frac{4}{\pi}\cos(\pi x) \\
f_2(x) =& \frac{4}{\pi}\left(\cos(\pi x) - \frac{1}{3}\cos(3\pi x)\right)
\end{aligned}
$$

$$f_3(x) = \frac{4}{\pi}\left(\cos(\pi x) - \frac{1}{3}\cos(3\pi x) + \frac{1}{5}\cos(5\pi x)\right)$$

$$\vdots$$

we see the sequence of continuous functions (f_n) converges to the non-continuous function f for each real number x. This didn't happen with Taylor series. The partial sums for a Taylor series were polynomials and hence continuous but what they converged to was continuous as well.

The difficulty is quite delicate and it took mathematicians a while to determine the problem. There are two very subtly different ways that a sequence of functions can converge: pointwise or uniformly. This distinction was touched upon by Niels Henrik Abel (1802-1829) in 1826 while studying the domain of convergence of a power series. However, the necessary formal definitions were not made explicit until Weierstrass did it in his 1841 paper *Zur Theorie der Potenzreihen (On the Theory of Power Series)*. This was published in his collected works in 1894.

It will be instructive to take a look at an argument that doesn't quite work before looking at the formal definitions we will need. In 1821 Augustin Cauchy "proved" that the infinite sum of continuous functions is continuous. Of course, it is obvious (to us) that this is not true because we've seen several counterexamples. But Cauchy, who was a first rate mathematician was so sure of the correctness of his argument that he included it in his textbook on analysis, *Cours d'analyse* (1821).

Problem 10.1.1 Find the flaw in the following "proof" that f is also continuous at a.

Suppose $f_1, f_2, f_3, f_4 \ldots$ are all continuous at a and that $\sum_{n=1}^{\infty} f_n = f$. Let $\varepsilon > 0$. Since f_n is continuous at a, we can choose $\delta_n > 0$ such that if $|x-a| < \delta_n$, then $|f_n(x) - f_n(a)| < \frac{\varepsilon}{2^n}$. Let $\delta = \inf(\delta_1, \delta_2, \delta_3, \ldots)$. If $|x-a| < \delta$ then

$$\begin{aligned}
|f(x) - f(a)| &= \left|\sum_{n=1}^{\infty} f_n(x) - \sum_{n=1}^{\infty} f_n(a)\right| \\
&= \left|\sum_{n=1}^{\infty} (f_n(x) - f_n(a))\right| \\
&\leq \sum_{n=1}^{\infty} |f_n(x) - f_n(a)| \\
&\leq \sum_{n=1}^{\infty} \frac{\varepsilon}{2^n} \\
&\leq \varepsilon \sum_{n=1}^{\infty} \frac{1}{2^n} \\
&= \varepsilon.
\end{aligned}$$

Thus f is continuous at a. □

Definition 10.1.2 Let S be a subset of the real number system and let $(f_n) = (f_1, f_2, f_3, \ldots)$ be a sequence of functions defined on S. Let f be a function defined on S as well. We say that (f_n) ***converges to f pointwise on S*** provided that for all $x \in S$, the sequence of real numbers $(f_n(x))$ converges to the number $f(x)$. In this case we write $f_n \overset{ptwise}{\longrightarrow} f$ on S. ◊

Symbolically, we have $f_n \overset{ptwise}{\longrightarrow} f$ on $S \Leftrightarrow \forall\, x \in S, \forall\, \varepsilon > 0, \exists\, N$ such that $(n > N \Rightarrow |f_n(x) - f(x)| < \varepsilon)$.

This is the type of convergence we have been observing to this point. By contrast we have the following new definition.

Definition 10.1.3 Let S be a subset of the real number system and let $(f_n) = (f_1, f_2, f_3, \ldots)$ be a sequence of functions defined on S. Let f be a function defined on S as well. We say that (f_n) ***converges to f uniformly on S*** provided $\forall\, \varepsilon > 0, \exists\, N$ such that $n > N \Rightarrow |f_n(x) - f(x)| < \varepsilon\ , \forall\, x \in S$.

In this case we write $f_n \overset{unif}{\longrightarrow} f$ on S. ♢

The difference between these two definitions is subtle. In pointwise convergence, we are given a fixed $x \in S$ and an $\varepsilon > 0$. Then the task is to find an N that works for that particular x and ε. In uniform convergence, one is given $\varepsilon > 0$ and must find a single N that works for that particular ε but also simultaneously (uniformly) for all $x \in S$. Clearly uniform convergence implies pointwise convergence as an N which works uniformly for all x, works for each individual x also. However the reverse is not true. This will become evident, but first consider the following example.

Problem 10.1.4 Let $0 < b < 1$ and consider the sequence of functions (f_n) defined on $[0, b]$ by $f_n(x) = x^n$. Use the definition to show that $f_n \overset{unif}{\longrightarrow} 0$ on $[0, b]$.

Hint.
$|x^n - 0| = x^n \leq b^n$. □

Uniform convergence is not only dependent on the sequence of functions but also on the set S. For example, the sequence $(f_n(x)) = (x^n)_{n=0}^{\infty}$ of Problem 10.1.4 does not converge uniformly on $[0, 1]$. We could use the negation of the definition to prove this, but instead, it will be a consequence of the following theorem.

Theorem 10.1.5 *Consider a sequence of functions (f_n) which are all continuous on an interval I. Suppose $f_n \overset{unif}{\longrightarrow} f$ on I. Then f must be continuous on I.*

Sketch of Proof. Let $a \in I$ and let $\varepsilon > 0$. The idea is to use uniform convergence to replace f with one of the known continuous functions f_n. Specifically, by uncancelling, we can write

$$\begin{aligned}|f(x) - f(a)| &= |f(x) - f_n(x) + f_n(x) - f_n(a) + f_n(a) - f(a)| \\ &\leq |f(x) - f_n(x)| + |f_n(x) - f_n(a)| + |f_n(a) - f(a)|\end{aligned}$$

If we choose n large enough, then we can make the first and last terms as small as we wish, noting that the uniform convergence makes the first term uniformly small for all x. Once we have a specific n, then we can use the continuity of f_n to find a $\delta > 0$ such that the middle term is small whenever x is within δ of a. ■

Problem 10.1.6 Provide a formal proof of Theorem 10.1.5 based on the above ideas. □

Problem 10.1.7 Consider the sequence of functions (f_n) defined on $[0, 1]$ by $f_n(x) = x^n$. Show that the sequence converges to the function

$$f(x) = \begin{cases} 0 & \text{if } x \in [0, 1) \\ 1 & \text{if } x = 1 \end{cases}$$

pointwise on $[0, 1]$, but not uniformly on $[0, 1]$. □

Notice that for the Fourier series at the beginning of this chapter,

$$f(x) = \frac{4}{\pi}\left(\cos\left(\frac{\pi}{2}x\right) - \frac{1}{3}\cos(3\pi x) + \frac{1}{5}\cos(5\pi x) - \frac{1}{7}\cos(7\pi x) + \cdots\right)$$

the convergence cannot be uniform on $(-\infty, \infty)$, as the function f is not continuous. This never happens with a power series, since they converge to continuous functions whenever they converge. We will also see that uniform convergence is what allows us to integrate and differentiate a power series term by term.

10.2 Uniform Convergence: Integrals and Derivatives

We saw in the previous section that if (f_n) is a sequence of continuous functions which converges uniformly to f on an interval, then f must be continuous on the interval as well. This was not necessarily true if the convergence was only pointwise, as we saw a sequence of continuous functions defined on $(-\infty, \infty)$ converging pointwise to a Fourier series that was not continuous on the real line. Uniform convergence guarantees some other nice properties as well.

Theorem 10.2.1 *Suppose f_n and f are integrable and $f_n \overset{unif}{\longrightarrow} f$ on $[a, b]$. Then*

$$\lim_{n\to\infty} \int_{x=a}^{b} f_n(x)\,\mathrm{d}x = \int_{x=a}^{b} f(x)\,\mathrm{d}x.$$

Problem 10.2.2 Prove Theorem 10.2.1.

Hint.
For $\varepsilon > 0$, we need to make $|f_n(x) - f(x)| < \frac{\varepsilon}{b-a}$, for all $x \in [a, b]$. □

Notice that this theorem is not true if the convergence is only pointwise, as illustrated by the following.

Problem 10.2.3 Consider the sequence of functions (f_n) given by

$$f_n(x) = \begin{cases} n & \text{if } x \in \left(0, \frac{1}{n}\right) \\ 0 & \text{otherwise} \end{cases}.$$

a Show that $f_n \overset{ptwise}{\longrightarrow} 0$ on $[0, 1]$, but $\lim_{n\to\infty} \int_{x=0}^{1} f_n(x)\,\mathrm{d}x \neq \int_{x=0}^{1} 0\,\mathrm{d}x$.

b Can the convergence be uniform? Explain.

□

Applying this result to power series we have the following.

Comment. Notice that we must explicitly assume uniform convergence. This is because we have not *yet* proved that power series actually do converge uniformly.

Corollary 10.2.4 *If $\sum_{n=0}^{\infty} a_n x^n$ converges uniformly to f on an interval containing 0 and x then $\int_{t=0}^{x} f(t)\,\mathrm{d}t = \sum_{n=0}^{\infty}\left(\frac{a_n}{n+1}x^{n+1}\right)$.*

Problem 10.2.5 Prove Corollary 10.2.4.

Hint.

Remember that

$$\sum_{n=0}^{\infty} f_n(x) = \lim_{N\to\infty} \sum_{n=0}^{N} f_n(x).$$

□

Surprisingly, the issue of term-by-term differentiation depends not on the uniform convergence of (f_n), but on the uniform convergence of (f_n'). More precisely, we have the following result.

Theorem 10.2.6 *Suppose for every $n \in \mathbb{N}$ f_n is differentiable, f_n' is continuous, $f_n \overset{ptwise}{\longrightarrow} f$, and $f_n' \overset{unif}{\longrightarrow} g$ on an interval, I. Then f is differentiable and $f' = g$ on I.*

Problem 10.2.7 Prove Theorem 10.2.6.

Hint.

Let a be an arbitrary fixed point in I and let $x \in I$. By the Fundamental Theorem of Calculus, we have

$$\int_{t=a}^{x} f_n'(t)\,\mathrm{d}t = f_n(x) - f_n(a).$$

Take the limit of both sides and differentiate with respect to x. □

As before, applying this to power series gives the following result.

Corollary 10.2.8 *If $\sum_{n=0}^{\infty} a_n x^n$ converges pointwise to f on an interval containing 0 and x and $\sum_{n=1}^{\infty} a_n n x^{n-1}$ converges uniformly on an interval containing 0 and x, then $f'(x) = \sum_{n=1}^{\infty} a_n n x^{n-1}$.*

Problem 10.2.9 Prove Corollary 10.2.8. □

The above results say that a power series can be differentiated and integrated term-by-term as long as the convergence is uniform. Fortunately it is, in general, true that when a power series converges the convergence of it and its integrated and differentiated series is also uniform (almost).

However we do not yet have all of the tools necessary to see this. To build these tools requires that we return briefly to our study, begun in Chapter 6, of the convergence of sequences.

10.2.1 Cauchy Sequences

Knowing that a sequence or a series converges and knowing what it converges to are typically two different matters. For example, we know that $\sum_{n=0}^{\infty} \frac{1}{n!}$and $\sum_{n=0}^{\infty} \frac{1}{n!\,n!}$ both converge. The first converges to e, which has meaning in other contexts. We don't know what the second one converges to, other than to say it converges to $\sum_{n=0}^{\infty} \frac{1}{n!\,n!}$. In fact, that question might not have much meaning without some other context in which $\sum_{n=0}^{\infty} \frac{1}{n!\,n!}$ arises naturally. Be that as it may, we need to look at the convergence of a series (or a sequence for that matter) without necessarily knowing what it might converge to. We make the following definition.

Definition 10.2.10 Let (s_n) be a sequence of real numbers. We say that $(s_n)$is a **Cauchy sequence** if for any $\varepsilon > 0$, there exists a real number N such that if $m, n > N$, then $|s_m - s_n| < \varepsilon$. ◊

Notice that this definition says that the terms in a Cauchy sequence get arbitrarily close to each other and that there is no reference to getting close to any particular fixed real number. Furthermore, you have already seen lots of examples of Cauchy sequences as illustrated by the following result.

Theorem 10.2.11 *Suppose (s_n) is a sequence of real numbers which converges to s. Then (s_n) is a Cauchy sequence.*

Intuitively, this result makes sense. If the terms in a sequence are getting arbitrarily close to s, then they should be getting arbitrarily close to each other. This is the basis of the proof.

Comment. But the converse isn't nearly as clear. In fact, it isn't true in the rational numbers.

Problem 10.2.12 Prove Theorem 10.2.11.

Hint.

$|s_m - s_n| = |s_m - s + s - s_n| \leq |s_m - s| + |s - s_n|$. □

So any convergent sequence is automatically Cauchy. For the real number system, the converse is also true and, in fact, is equivalent to any of our completeness axioms: the NIP, the Bolzano-Weierstrass Theorem, or the LUB Property. Thus, this could have been taken as our completeness axiom and we could have used it to prove the others. One of the most convenient ways to prove this converse is to use the Bolzano-Weierstrass Theorem. To do that, we must first show that a Cauchy sequence must be bounded. This result is reminiscent of the fact that a convergent sequence is bounded (Lemma 6.2.7 of Chapter 6) and the proof is very similar.

Lemma 10.2.13 *Suppose (s_n) is a Cauchy sequence. Then there exists $B > 0$ such that $|s_n| \leq B$ for all n.*

Problem 10.2.14 Prove Lemma 10.2.13.

Hint.

This is similar to problem 6.2.8 of Chapter 6. There exists N such that if $m, n > N$then $|s_n - s_m| < 1$. Choose a fixed $m > N$ and let $B = \max\left(|s_1|, |s_2|, \ldots, \left|s_{\lceil N \rceil}\right|, |s_m| + 1\right)$. □

Theorem 10.2.15 ***Cauchy sequences converge***

Suppose (s_n) is a Cauchy sequence of real numbers. There exists a real number s such that $\lim_{n\to\infty} s_n = s$.

Sketch of Proof. We know that (s_n) is bounded, so by the Bolzano-Weierstrass Theorem, it has a convergent subsequence (s_{n_k}) converging to some real number s. We have $|s_n - s| = |s_n - s_{n_k} + s_{n_k} - s| \leq |s_n - s_{n_k}| + |s_{n_k} - s|$. If we choose n and n_k large enough, we should be able to make each term arbitrarily small. ■

Problem 10.2.16 Provide a formal proof of Theorem 10.2.15. □

From Theorem 10.2.11 we see that every Cauchy sequence converges in $\mathbb{R}$. Moreover the proof of this fact depends on the Bolzano-Weierstrass Theorem which, as we have seen, is equivalent to our completeness axiom, the Nested Interval Property. What this means is that if there is a Cauchy sequence which does not converge then the NIP is not true. A natural question to ask is if every Cauchy sequence converges does the NIP follow? That is, is the convergence of Cauchy sequences also equivalent to our completeness axiom? The following theorem shows that the answer is yes.

Theorem 10.2.17 *Suppose every Cauchy sequence converges. Then the Nested Interval Property is true.*

Problem 10.2.18 Prove Theorem 10.2.17.

Hint.

If we start with two sequences (x_n) and (y_n), satisfying all of the conditions of the NIP, you should be able to show that these are both Cauchy sequences.

□

Problems 10.2.16 and Problem 10.2.18 tell us that the following are equivalent: the Nested Interval Property, the Bolzano-Weierstrass Theorem, the Least Upper Bound Property, and the convergence of Cauchy sequences. Thus any one of these could have been taken as the completeness axiom of the real number system and then used to prove the each of the others as a theorem according to the following dependency graph:

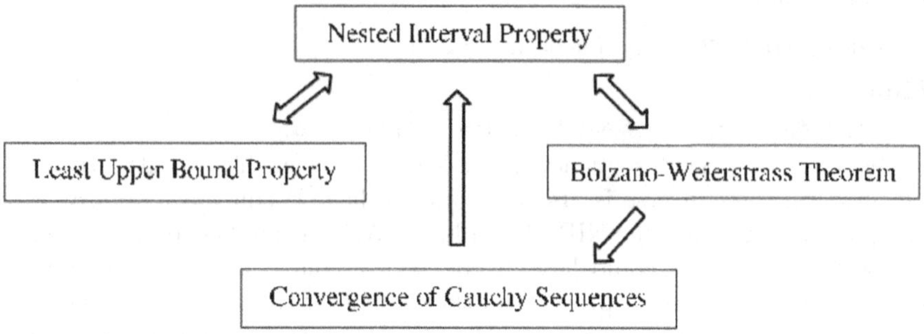

Since we can get from any node on the graph to any other, simply by following the implications (indicated with arrows), any one of these statements is logically equivalent to each of the others.

Problem 10.2.19 Since the convergence of Cauchy sequences can be taken as the completeness axiom for the real number system, it does not hold for the rational number system. Give an example of a Cauchy sequence of rational numbers which does not converge to a rational number. □

If we apply the above ideas to series we obtain the following important result, which will provide the basis for our investigation of power series.

Theorem 10.2.20 Cauchy Criterion. *The series $\sum_{k=0}^{\infty} a_k$ converges if and only if $\forall\ \varepsilon > 0,\ \exists\ N$ such that if $m > n > N$ then $|\sum_{k=n+1}^{m} a_k| < \varepsilon$.*

Problem 10.2.21 Prove the Cauchy criterion. □

At this point several of the tests for convergence that you probably learned in calculus are easily proved. For example:

Problem 10.2.22 *The nth Term Test*
Show that if $\sum_{n=1}^{\infty} a_n$ converges then $\lim_{n\to\infty} a_n = 0$. □

Problem 10.2.23 The Strong Cauchy Criterion. Show that $\sum_{k=1}^{\infty} a_k$ converges if and only if $\lim_{n\to\infty} \sum_{k=n+1}^{\infty} a_k = 0$.

Hint.
The hardest part of this problem is recognizing that it is really about the limit of a sequence as in Chapter 6. □

You may also recall the Comparison Test from studying series in calculus: suppose $0 \leq a_n \leq b_n$, if $\sum b_n$ converges then $\sum a_n$ converges. This result follows from the fact that the partial sums of $\sum a_n$ form an increasing sequence which is bounded above by $\sum b_n$. (See Corollary 9.4.5 of Chapter 9.) The Cauchy Criterion allows us to extend this to the case where the terms a_n could be negative as well. This can be seen in the following theorem.

Theorem 10.2.24 *Comparison Test*
Suppose $|a_n| \leq b_n$ for all n. If $\sum b_n$ converges then $\sum a_n$ also converges.

Problem 10.2.25 Prove Theorem 10.2.24.

Hint.
Use the Cauchy criterion with the fact that $\left|\sum_{k=n+1}^{m} a_k\right| \le \sum_{k=n+1}^{m} |a_k|$. □

The following definition is of marked importance in the study of series.

Definition 10.2.26 ***Absolute Convergence***
Given a series $\sum a_n$, the series $\sum |a_n|$ is called the **absolute series** of $\sum a_n$ and if $\sum |a_n|$ converges then we say that $\sum a_n$ **converges absolutely.** ◊

The significance of this definition comes from the following result.

Corollary 10.2.27 *If $\sum a_n$ converges absolutely, then $\sum a_n$ converges.*

Problem 10.2.28 Show that Corollary 10.2.27 is a direct consequence of Theorem 10.2.24. □

Problem 10.2.29 If $\sum_{n=0}^{\infty} |a_n| = s$, then does it follow that $s = \left|\sum_{n=0}^{\infty} a_n\right|$? Justify your answer. What can be said? □

The converse of Corollary 10.2.27 is not true as evidenced by the series $\sum_{n=0}^{\infty} \frac{(-1)^n}{n+1}$. As we noted in Chapter 4, this series converges to ln 2. However, its absolute series is the Harmonic Series which diverges. Any such series which converges, but not absolutely, is said to ***converge conditionally.*** Recall also that in Chapter 4, we showed that we could rearrange the terms of the series $\sum_{n=0}^{\infty} \frac{(-1)^n}{n+1}$ to make it converge to any number we wished. We noted further that all rearrangements of the series $\sum_{n=0}^{\infty} \frac{(-1)^n}{(n+1)^2}$ converged to the same value. The difference between the two series is that the latter converges absolutely whereas the former does not. Specifically, we have the following result.

Theorem 10.2.30 *Suppose $\sum a_n$ converges absolutely and let $s = \sum_{n=0}^{\infty} a_n$. Then any rearrangement of $\sum a_n$ must converge to s.*
Sketch of Proof. We will first show that this result is true in the case where $a_n \ge 0$. If $\sum b_n$ represents a rearrangement of $\sum a_n$, then notice that the sequence of partial sums $\left(\sum_{k=0}^{n} b_k\right)_{n=0}^{\infty}$is an increasing sequence which is bounded by s. By Corollary 9.4.5 of Chapter 9, this sequence must converge to some number t and $t \le s$. Furthermore $\sum a_n$ is also a rearrangement of $\sum b_n$. Thus the result holds for this special case. (Why?) For the general case, notice that $a_n = \frac{|a_n|+a_n}{2} - \frac{|a_n|-a_n}{2}$ and that $\sum \frac{|a_n|+a_n}{2}$ and $\sum \frac{|a_n|-a_n}{2}$ are both convergent series with nonnegative terms. By the special case $\sum \frac{|b_n|+b_n}{2} = \sum \frac{|a_n|+a_n}{2}$ and $\sum \frac{|b_n|-b_n}{2} = \sum \frac{|a_n|-a_n}{2}$. ■

Problem 10.2.31 Fill in the details and provide a formal proof of Theorem 10.2.30. □

10.3 Radius of Convergence of a Power Series

We've developed enough machinery to look at the convergence of power series. The fundamental result is the following theorem due to Abel.

Theorem 10.3.1 *Suppose $\sum_{n=0}^{\infty} a_n c^n$ converges for some nonzero real number c. Then $\sum_{n=0}^{\infty} a_n x^n$ converges absolutely for all x such that $|x| < |c|$.*

To prove Theorem 10.3.1 first note that by Problem 10.2.22, $\lim_{n\to\infty} a_n c^n = 0$. Thus $(a_n c^n)$ is a bounded sequence. Let B be a bound: $|a_n c^n| \le B$. Then

$$|a_n x^n| = \left|a_n c^n \cdot \left(\frac{x}{c}\right)^n\right| \le B\left|\frac{x}{c}\right|^n.$$

We can now use the comparison test.

Problem 10.3.2 Prove Theorem 10.3.1. □

Corollary 10.3.3 *Suppose* $\sum_{n=0}^{\infty} a_n c^n$ *diverges for some real number* c. *Then* $\sum_{n=0}^{\infty} a_n x^n$ *diverges for all* x *such that* $|x| > |c|$.

Problem 10.3.4 Prove Corollary 10.3.3. □

As a result of Theorem 10.3.1 and Corollary 10.3.3, we have the following: either $\sum_{n=0}^{\infty} a_n x^n$ converges absolutely for all x or there exists some nonnegative real number r such that $\sum_{n=0}^{\infty} a_n x^n$ converges absolutely when $|x| < r$ and diverges when $|x| > r$. In the latter case, we call r the *radius of convergence* of the power series $\sum_{n=0}^{\infty} a_n x^n$. In the former case, we say that the radius of convergence of $\sum_{n=0}^{\infty} a_n x^n$ is ∞. Though we can say that $\sum_{n=0}^{\infty} a_n x^n$ converges absolutely when $|x| < r$, we cannot say that the convergence is uniform. However, we can come close. We can show that the convergence is uniform for $|x| \le b < r$. To see this we will use the following result

Theorem 10.3.5 ***The Weierstrass-M Test***

Let $(f_n)_{n=1}^{\infty}$ *be a sequence of functions defined on* $S \subseteq \mathbb{R}$ *and suppose that* $(M_n)_{n=1}^{\infty}$ *is a sequence of nonnegative real numbers such that*

$$|f_n(x)| \le M_n, \ \forall x \in S, \ n = 1, 2, 3, \ldots.$$

If $\sum_{n=1}^{\infty} M_n$ *converges then* $\sum_{n=1}^{\infty} f_n(x)$ *converges uniformly on* S *to some function (which we will denote by* $f(x)$*).*

Sketch of Proof. Since the crucial feature of the theorem is the function $f(x)$ that our series converges to, our plan of attack is to first define $f(x)$ and then show that our series, $\sum_{n=1}^{\infty} f_n(x)$, converges to it uniformly.

First observe that for any $x \in S$, $\sum_{n=1}^{\infty} f_n(x)$ converges by the Comparison Test (in fact it converges absolutely) to some number we will denote by $f(x)$. This actually defines the function $f(x)$ for all $x \in S$. It follows that $\sum_{n=1}^{\infty} f_n(x)$ converges pointwise to $f(x)$.

Next, let $\varepsilon > 0$ be given. Notice that since $\sum_{n=1}^{\infty} M_n$ converges, say to M, then there is a real number, N, such that if $n > N$, then

$$\sum_{k=n+1}^{\infty} M_k = \left|\sum_{k=n+1}^{\infty} M_k\right| = \left|M - \sum_{k=1}^{n} M_k\right| < \varepsilon.$$

You should be able to use this to show that if $n > N$, then

$$\left|f(x) - \sum_{k=1}^{n} f_k(x)\right| < \varepsilon, \ \forall x \in S.$$

■

Problem 10.3.6 Use the ideas above to provide a formal proof of Theorem 10.3.5. □

Problem 10.3.7

(a) Referring back to equation (II.1), show that the Fourier series

$$\sum_{k=0}^{\infty} \frac{(-1)^k}{(2k+1)^2} \sin\left((2k+1)\pi x\right)$$

converges uniformly on $\mathbb{R}$.

(b) Does its differentiated series converge uniformly on $\mathbb{R}$? Explain.

□

Problem 10.3.8 Observe that for all $x \in [-1,1]$ $|x| \leq 1$. Identify which of the following series converges pointwise and which converges uniformly on the interval $[-1,1]$. In every case identify the limit function.

(a) $\displaystyle\sum_{n=1}^{\infty}\left(x^n - x^{n-1}\right)$

(b) $\displaystyle\sum_{n=1}^{\infty}\frac{\left(x^n - x^{n-1}\right)}{n}$

(c) $\sum_{n=1}^{\infty}\frac{\left(x^n - x^{n-1}\right)}{n^2}$

□

Using the Weierstrass-M test, we can prove the following result.

Theorem 10.3.9 *Suppose $\sum_{n=0}^{\infty} a_n x^n$ has radius of convergence r (where r could be ∞ as well). Let b be any nonnegative real number with $b < r$. Then $\sum_{n=0}^{\infty} a_n x^n$ converges uniformly on $[-b,b]$.*

Problem 10.3.10 Prove Theorem 10.3.9.

Hint.

We know that $\sum_{n=0}^{\infty} |a_n b^n|$ converges. This should be all set for the Weierstrass-M test. □

To finish the story on differentiating and integrating power series, all we need to do is show that the power series, its integrated series, and its differentiated series all have the same radius of convergence. You might not realize it, but we already know that the integrated series has a radius of convergence at least as big as the radius of convergence of the original series. Specifically, suppose $f(x) = \sum_{n=0}^{\infty} a_n x^n$has a radius of convergence r and let $|x| < r$. We know that $\sum_{n=0}^{\infty} a_n x^n$ converges uniformly on an interval containing 0and x, and so by Corollary 10.2.4, $\int_{t=0}^{x} f(t)\,\mathrm{d}t = \sum_{n=0}^{\infty}\left(\frac{a_n}{n+1}x^{n+1}\right)$. In other words, the integrated series converges for any x with $|x| < r$. This says that the radius of convergence of the integated series must be at least r.

To show that the radii of convergence are the same, all we need to show is that the radius of convergence of the differentiated series is at least as big as r as well. Indeed, since the differentiated series of the integrated series is the original, then this would say that the original series and the integrated series have the same radii of convergence. Putting the differentiated series into the role of the original series, the original series is now the integrated series and so these would have the same radii of convergence as well. With this in mind, we want to show that if $|x| < r$, then $\sum_{n=0}^{\infty} a_n n x^{n-1}$ converges. The strategy is to mimic what we did in Theorem 10.3.1, where we essentially compared our series with a converging geometric series. Only this time we need to start with the differentiated geometric series.

Problem 10.3.11 Show that $\sum_{n=1}^{\infty} n x^{n-1}$ converges for $|x| < 1$.

Hint.

We know that $\sum_{k=0}^{n} x^k = \frac{x^{n+1}-1}{x-1}$. Differentiate both sides and take the limit as n approaches infinity. □

Theorem 10.3.12 *Suppose $\sum_{n=0}^{\infty} a_n x^n$ has a radius of convergence r and let $|x| < r$. Then $\sum_{n=1}^{\infty} a_n n x^{n-1}$ converges.*

Problem 10.3.13 Prove Theorem 10.3.12.

Hint.
Let b be a number with $|x| < b < r$ and consider $\left|a_n n x^{n-1}\right| = \left|a_n b^n \cdot \frac{1}{b} \cdot n\left(\frac{x}{b}\right)^{n-1}\right|$. You should be able to use the Comparison Test and Problem 10.3.11. □

10.4 Boundary Issues and Abel's Theorem

Summarizing our results, we see that any power series $\sum a_n x^n$ has a radius of convergence r such that $\sum a_n x^n$ converges absolutely when $|x| < r$ and diverges when $|x| > r$. Furthermore, the convergence is uniform on any closed interval $[-b, b] \subset (-r, r)$ which tells us that whatever the power series converges to must be a continuous function on $(-r, r)$. Lastly, if $f(x) = \sum_{n=0}^{\infty} a_n x^n$ for $x \in (-r, r)$, then $f'(x) = \sum_{n=1}^{\infty} a_n n x^{n-1}$ for $x \in (-r, r)$ and $\int_{t=0}^{x} f(t)\,\mathrm{d}t = \sum_{n=0}^{\infty} a_n \frac{x^{n+1}}{n+1}$ for $x \in (-r, r)$.

Thus power series are very well behaved within their interval of convergence, and our cavalier approach from Chapter 3 is justified, ***EXCEPT*** for one issue. If you go back to Problem 3.2.10 of Chapter 3, you see that we used the geometric series to obtain the series, $\arctan x = \sum_{n=0}^{\infty}(-1)^n \frac{1}{2n+1} x^{2n+1}$. We substituted $x = 1$ into this to obtain $\frac{\pi}{4} = \sum_{n=0}^{\infty}(-1)^n \frac{1}{2n+1}$. Unfortunately, our integration was only guaranteed on a closed subinterval of the interval $(-1, 1)$ where the convergence was uniform and we substituted in $x = 1$. We "danced on the boundary" in other places as well, including when we said that

$$\frac{\pi}{4} = \int_{x=0}^{1} \sqrt{1 - x^2}\,\mathrm{d}x = 1 + \sum_{n=1}^{\infty} \left(\frac{\prod_{j=0}^{n-1}\left(\frac{1}{2} - j\right)}{n!}\right)\left(\frac{(-1)^n}{2n+1}\right).$$

The fact is that for a power series $\sum a_n x^n$ with radius of convergence r, we know what happens when $|x| < r$ and when $|x| > r$. But we've never talked about what happens when $|x| = r$. That is because there is no systematic approach to this boundary problem. For example, consider the three series

$$\sum_{n=0}^{\infty} x^n, \sum_{n=0}^{\infty} \frac{x^{n+1}}{n+1}, \sum_{n=0}^{\infty} \frac{x^{n+2}}{(n+1)(n+2)}.$$

They are all related in that we started with the geometric series and integrated twice, thus they all have radius of convergence equal to 1. Their behavior on the boundary, i.e., when $x = \pm 1$, is another story. The first series diverges when $x = \pm 1$, the third series converges when $x = \pm 1$. The second series converges when $x = -1$ and diverges when $x = 1$.

Even with the unpredictability of a power series at the endpoints of its interval of convergence, the Weierstrass-M test does give us some hope of uniform convergence.

Problem 10.4.1 Suppose the power series $\sum a_n x^n$ has radius of convergence r and the series $\sum a_n r^n$ converges absolutely. Then $\sum a_n x^n$ converges uniformly on $[-r, r]$.

Hint.
For $|x| \leq r$, $|a_n x^n| \leq |a_n r^n|$. □

Unfortunately, this result doesn't apply to the integrals we mentioned as the convergence at the endpoints is not absolute. Nonetheless, the integrations we performed in Chapter 3 are still legitimate. This is due to the following theorem by Abel which extends uniform convergence to the endpoints of the interval of convergence even if the convergence at an endpoint is only conditional. Abel did not use the term uniform convergence, as it hadn't been defined yet, but the ideas involved are his.

Theorem 10.4.2 ***Abel's Theorem***

Suppose the power series $\sum a_n x^n$ has radius of convergence r and the series $\sum a_n r^n$ converges. Then $\sum a_n x^n$ converges uniformly on $[0, r]$.

The proof of this is not intuitive, but involves a clever technique known as Abel's Partial Summation Formula.

Lemma 10.4.3 *Let*

$$a_1, a_2, \ldots, a_n, b_1, b_2, \ldots, b_n$$

be real numbers and let $A_m = \sum_{k=1}^{m} a_k$. Then

$$a_1 b_1 + a_2 b_2 + \cdots + a_n b_n = \sum_{j=1}^{n-1} A_j \left(b_j - b_{j+1}\right) + A_n b_n.$$

Problem 10.4.4 Prove Lemma 10.4.3.

Hint.

For $j > 1$, $a_j = A_j - A_{j-1}$. □

Lemma 10.4.5 ***Abel's Lemma***

Let $a_1, a_2, \ldots, a_n, b_1, b_2, \ldots, b_n$ be real numbers with $b_1 \geq b_2 \geq \ldots \geq b_n \geq 0$ and let $A_m = \sum_{k=1}^{m} a_k$. Suppose $|A_m| \leq B$ for all m. Then $\left|\sum_{j=1}^{n} a_j b_j\right| \leq B \cdot b_1$.

Problem 10.4.6 Prove Lemma 10.4.5. □

Problem 10.4.7 Prove Theorem 10.4.2.

Hint.

Let $\epsilon > 0$. Since $\sum_{n=0}^{\infty} a_n r^n$ converges then by the Cauchy Criterion, there exists N such that if $m > n > N$ then $\left|\sum_{k=n+1}^{m} a_k r^k\right| < \frac{\epsilon}{2}$. Let $0 \leq x \leq r$. By Lemma 10.4.5,

$$\left|\sum_{k=n+1}^{m} a_k x^k\right| = \left|\sum_{k=n+1}^{m} a_k r^k \left(\frac{x}{r}\right)^k\right| \leq \left(\frac{\epsilon}{2}\right)\left(\frac{x}{r}\right)^{n+1} \leq \frac{\epsilon}{2}.$$

Thus for $0 \leq x \leq r$, $n > N$,

$$\left|\sum_{k=n+1}^{\infty} a_k x^k\right| = \lim_{m \to \infty} \left|\sum_{k=n+1}^{m} a_k x^k\right| \leq \frac{\epsilon}{2} < \epsilon.]$$

□

Corollary 10.4.8 *Suppose the power series $\sum a_n x^n$ has radius of convergence r and the series $\sum a_n (-r)^n$ converges. Then $\sum a_n x^n$ converges uniformly on $[-r, 0]$.*

Problem 10.4.9 Prove Corollary 10.4.8.

Hint.

Consider $\sum a_n (-x)^n$. □

Chapter 11

Back to the Real Numbers

As we have seen, when they converge, power series are very well behaved and Fourier (trigonometric) series are not necessarily. The fact that trigonometric series were so interesting made them a lightning rod for mathematical study in the late nineteenth century.

For example, consider the question of uniqueness. We saw in Chapter 7 that if a function could be represented by a power series, then that series must be the Taylor series. More precisely, if

$$f(x) = \sum_{n=0}^{\infty} a_n (x-a)^n, \text{ then } a_n = \frac{f^{(n)}(a)}{n!}.$$

But what can be said about the uniqueness of a trigonometric series? If we can represent a function f as a general trigonometric series

$$f(x) = \sum_{n=0}^{\infty} (a_n \cos n\pi x + b_n \sin n\pi x),$$

then must this be the Fourier series with the coefficients as determined by Fourier?

For example, if $\sum_{n=0}^{\infty} (a_n \cos n\pi x + b_n \sin n\pi x)$ converges to f uniformly on the interval $(0, 1)$, then because of the uniform convergence, Fourier's term-by-term integration which we saw in Part II is perfectly legitimate and the coefficients are, of necessity, the coefficients he computed. However we have seen that the convergence of a Fourier series need not be uniform. This does not mean that we cannot integrate term-by-term, but it does say that we can't be sure that term-by-term integration of a Fourier series will yield the integral of the associated function.

This led to a generalization of the integral by Henri Lebesgue in 1905. Lebesgue's profound work settled the issue of whether or not a bounded pointwise converging trigonometric series is the Fourier series of a function, but we will not go in this direction. We will instead focus on work that Georg Cantor (1845-1918) did in the years just prior. Cantor's work was also profound and had far reaching implications in modern mathematics. It also leads to some very weird conclusions.

Comment. 'Weird' does not mean false. It simply means that some of Cantor's results can be hard to accept, even after you have seen the proof and verified its validity.

To begin, let's suppress the underlying function and suppose we have

$$\sum_{n=0}^{\infty}(a_n \cos n\pi x + b_n \sin n\pi x) = \sum_{n=0}^{\infty}(a'_n \cos n\pi x + b'_n \sin n\pi x).$$

We ask: If these two series are equal must it be true that $a_n = a'_n$ and $b_n = b'_n$? We can reformulate this uniqueness question as follows: Suppose

$$\sum_{n=0}^{\infty}((a_n - a'_n)\cos n\pi x + (b_n - b'_n)\sin n\pi x) = 0.$$

If we let $c_n = a_n - a'_n$ and $d_n = b_n - b'_n$, then the question becomes: If $\sum_{n=0}^{\infty}(c_n \cos n\pi x + d_n \sin n\pi x) = 0$, then will $c_n = d_n = 0$? It certainly seems reasonable to suppose so, but at this point we have enough experience with infinite sums to know that we need to be *very* careful about relying on the intuition we have honed on finite sums.

The answer to this seemingly basic question leads to some very profound results. In particular, answering this question led the mathematician Cantor to study the makeup of the real number system. This in turn opened the door to the modern view of mathematics of the twentieth century. In particular, Cantor proved the following result in 1871 ([6], p. 305).

Theorem 11.0.1 ***Cantor***
If the trigonometric series

$$\sum_{n=0}^{\infty}(c_n \cos n\pi x + d_n \sin n\pi x) = 0,$$

"with the exception of certain values of x," then all of its coefficients vanish.

In his attempts to nail down precisely which "certain values" could be exceptional, Cantor was led to examine the nature of subsets of real numbers and ultimately to give a precise definition of the concept of infinite sets and to define an arithmetic of "infinite numbers."

Transfinite Numbers. Actually, he called them **transfinite numbers** because, by definition, numbers are finite.

As a first step toward identifying those "certain values," Cantor proved the following theorem, which we will state but not prove.

Theorem 11.0.2 ***Cantor, (1870)***
If the trigonometric series

$$\sum_{n=0}^{\infty}(c_n \cos n\pi x + d_n \sin n\pi x) = 0,$$

for all $x \in \mathbb{R}$ then all of its coefficients vanish.

He then extended this to the following:

Theorem 11.0.3 ***Cantor, (1871)***
If the trigonometric series

$$\sum_{n=0}^{\infty}(c_n \cos n\pi x + d_n \sin n\pi x) = 0,$$

for all but finitely many $x \in \mathbb{R}$ then all of its coefficients vanish.

Figure 11.0.4 Georg Cantor[18]

Observe that this is not a trivial generalization. Although the exceptional points are constrained to be finite in number, this number could still be extraordinarily large. That is, even if the series given above differed from zero on $10^{10^{100000}}$ distinct points in the interval $(0, 10^{-10^{100000}})$ the coefficients still vanish. This remains true even if at each of these $10^{10^{100000}}$ points the series converges to $10^{10^{100000}}$. This is truly remarkable when you think of it this way.

At this point Cantor became more interested in these exceptional points than in the Fourier series problem that he'd started with. The next task he set for himself was to see just how general the set of exceptional points could be. Following Cantor's lead we make the following definitions.

Definition 11.0.5 Let $S \subseteq \mathbb{R}$ and let a be a real number. We say that a is a **limit point** (or an **accumulation point**) of S if there is a sequence (a_n) with $a_n \in S - \{a\}$ which converges to a. ◊

Problem 11.0.6 Let $S \subseteq \mathbb{R}$ and let a be a real number. Prove that a is a limit point of S if and only if for every $\varepsilon > 0$ the intersection

$$(a - \varepsilon, a + \varepsilon) \cap S - \{a\} \neq \emptyset.$$

□

The following definition gets to the heart of the matter.

Definition 11.0.7 Let $S \subseteq \mathbb{R}$. The set of all limit points of S is called the **derived set** of S. The derived set is denoted S'. ◊

Don't confuse the derived set of a set with the derivative of a function. They are completely different objects despite the similarity of both the language and notation. The only thing that they have in common is that they were somehow "derived" from something else.

Problem 11.0.8 Determine the derived set, S', of each of the following sets.

a $S = \left\{\frac{1}{1}, \frac{1}{2}, \frac{1}{3}, \ldots\right\}$

b $S = \left\{0, \frac{1}{1}, \frac{1}{2}, \frac{1}{3}, \ldots\right\}$

c $S = (0, 1]$

d $S = [0, 1/2) \cup (1/2, 1]$

[18] mathshistory.st-andrews.ac.uk/Biographies/Cantor/

e $S = \mathbb{Q}$

f $S = \mathbb{R} - \mathbb{Q}$

g $S = \mathbb{Z}$

h Any finite set S.

□

Problem 11.0.9 Let $S \subseteq \mathbb{R}$.

a Prove that $(S')' \subseteq S'$.

b Give an example where these two sets are equal.

c Give an example where these two sets are not equal.

□

The notion of the derived set forms the foundation of Cantor's exceptional set of values. Specifically, let S again be a set of real numbers and consider the following sequence of sets:

$$S' \supseteq (S')' \supseteq \left((S')'\right)' \supseteq \cdots.$$

Cantor showed that if, at some point, one of these derived sets is empty, then the uniqueness property still holds. Specifically, we have:

Theorem 11.0.10 ***Cantor, (1871)***
Let S be a subset of the real numbers with the property that one of its derived sets is empty. Then if the trigonometric series $\sum_{n=0}^{\infty} (c_n \cos n\pi x + d_n \sin n\pi x)$ is zero for all $x \in \mathbb{R} - S$, then all of the coefficients of the series vanish.

11.1 Infinite Sets

The following theorem follows directly from our previous work with the NIP and will be very handy later. It basically says that a sequence of nested closed intervals will still have a non-empty intersection even if their lengths do not converge to zero as in the NIP.

Theorem 11.1.1 *Let $([a_n, b_n])_{n=1}^{\infty}$ be a sequence of nested intervals such that $\lim_{n\to\infty} |b_n - a_n| > 0$. Then there is at least one $c \in \mathbb{R}$ such that $c \in [a_n, b_n]$ for all $n \in \mathbb{N}$.*

Proof. By Corollary 9.4.5 of Chapter 9, we know that a bounded increasing sequence such as (a_n) converges, say to c. Since $a_n \leq a_m \leq b_n$ for $m > n$ and $\lim_{m\to\infty} a_m = c$, then for any fixed n, $a_n \leq c \leq b_n$. This says $c \in [a_n, b_n]$ for all $n \in \mathbb{N}$. ■

Problem 11.1.2 Suppose $\lim_{n\to\infty} |b_n - a_n| > 0$. Show that there are at least two points, c and d, such that $c \in [a_n, b_n]$ and $d \in [a_n, b_n]$ for all $n \in \mathbb{N}$. □

Our next theorem says that in a certain, very technical sense there are more real numbers than there are counting numbers [3]. This probably does not seem terribly significant. After all, there are real numbers which are not counting numbers. What will make this so startling is that the same cannot be said about all sets which strictly contain the counting numbers. We will get into the details of this after the theorem is proved.

Theorem 11.1.3 ***Cantor, (1874)***
Let $S = (s_n)_{n=1}^{\infty}$ be a sequence of real numbers. There is a real number c, which is not in S.

Notation. *To streamline things, we are abusing notation here as we are letting S denote both the sequence (which is ordered) and the underlying (unordered) set of entries in the sequence.*

Proof. For the sake of obtaining a contradiction assume that the sequence S contains every real number; that is, $S = \mathbb{R}$. As usual we will build a sequence of nested intervals $([x_i, y_i])_{i=1}^{\infty}$.
Let x_1 be the smaller of the first two distinct elements of S, let y_1 be the larger and take $[x_1, y_1]$ to be the first interval.
Next we assume that $[x_{n-1}, y_{n-1}]$ has been constructed and build $[x_n, y_n]$ as follows. Observe that there are infinitely many elements of S in (x_{n-1}, y_{n-1}) since $S = \mathbb{R}$. Let s_m and s_k be the *first* two distinct elements of S such that

$$s_m, s_k \in (x_{n-1}, y_{n-1}).$$

Take x_n to be the smaller and y_n to be the larger of s_m and s_k. Then $[x_n, y_n]$ is the nth interval.
From the way we constructed them it is clear that

$$[x_1, y_1] \supseteq [x_2, y_2] \supseteq [x_3, y_3] \supseteq \dots.$$

Therefore by Theorem 11.1.1 there is a real number, say c, such that

$$c \in [x_n, y_n] \text{ for all } n \in \mathbb{N}.$$

In fact, since $x_1 < x_2 < x_3 \ldots < y_3 < y_2 < y_1$ it is clear that

$$x_n < c < y_n, \ \forall n. \tag{III.1}$$

We will show that c is the number we seek. That the inequalities in formula (III.1) are strict will play a crucial role.
To see that $c \notin S$ we suppose that $c \in S$ and derive a contradiction.
So, suppose that $c = s_p$ for some $p \in \mathbb{N}$. Then only $\{s_1, s_2, \ldots, s_{p-1}\}$ appear before s_p in the sequence S. Since each x_n is taken from S it follows that only finitely many elements of the sequence (x_n) appear before $s_p = c$ in the sequence as well.
Let x_l be the *last* element of (x_n) which appears before $c = s_p$ in the sequence and consider x_{l+1}. The way it was constructed, x_{l+1} was one of the first two distinct terms in the sequence S strictly between x_l and y_l, the other being y_{l+1}. Since x_{l+1} does not appear before $c = s_p$ in the sequence and $x_l < c < y_l$, it follows that either $c = x_{l+1}$ or $c = y_{l+1}$. However, this gives us a contradiction as we know from formula (III.1) that $x_{l+1} < c < y_{l+1}$.
Thus c is not an element of S. ■

So how does this theorem show that there are "more" real numbers than counting numbers? Before we address that question we need to be very careful about the meaning of the word 'more' when we're talking about infinite sets.

First let's consider two finite sets, say $A = \{\alpha, \beta, \gamma, \delta\}$ and $B = \{a, b, c, d, e\}$. How do we know that B is the bigger set? (It obviously is.) Clearly we can just count the number of elements in both A and B. Since $|A| = 4$ and $|B| = 5$ and $4 < 5$ B is clearly bigger. But we're looking for a way to determine the relative size of two sets without counting them because we have no way of counting the number of elements of an infinite set. Indeed, it isn't even clear what the phrase "the number of elements" might *mean* when applied to the elements of an infinite set.

When we count the number of elements in a finite set what we're really doing is matching up the elements of the set with a set of consecutive positive integers, starting at 1. Thus since

$$1 \leftrightarrow \alpha$$
$$2 \leftrightarrow \beta$$
$$3 \leftrightarrow \gamma$$
$$4 \leftrightarrow \delta$$

we see that $|A| = 4$. Moreover, the order of the match-up is unimportant. Thus since

$$2 \leftrightarrow e$$
$$3 \leftrightarrow a$$
$$5 \leftrightarrow b$$
$$4 \leftrightarrow d$$
$$1 \leftrightarrow c$$

it is clear that the elements of B and the set $\{1, 2, 3, 4, 5\}$ can be matched up as well. And it doesn't matter what order either set is in. They both have 5 elements.

Such a match-up is called a one-to-one correspondence. In general, if two sets can be put in one-to-one correspondence then they are the same "size." Of course the word "size" has lots of connotations that will begin to get in the way when we talk about infinite sets, so instead we will say that the two sets have the same *cardinality*. Speaking loosely, this just means that they are the same size.

More precisely, if a given set S can be put in one-to-one correspondence with a finite set of consecutive integers, say $\{1, 2, 3, \ldots, N\}$, then we say that the cardinality of the set is N. But this just means that both sets have the same cardinality. It is this notion of one-to-one correspondence, along with the next two definitions, which will allow us to compare the sizes (cardinalities) of infinite sets.

Definition 11.1.4 Any set which can be put into one-to-one correspondence with $\mathbb{N} = \{1, 2, 3, \ldots\}$ is called a **countably infinite** set. Any set which is either finite or countably infinite is said to be **countable**. ◊

Since $\mathbb{N}$ is an infinite set, we have no symbol to designate its cardinality so we have to invent one. The symbol used by Cantor and adopted by mathematicians ever since is $\aleph_0$. Thus the cardinality of any countably infinite set is $\aleph_0$.

Notation. $\aleph$ is the first letter of the Hebrew alphabet and is pronounced "aleph." $\aleph_0$ is pronounced "aleph null."

We have already given the following definition informally. We include it formally here for later reference.

Definition 11.1.5 If two sets can be put into one-to-one correspondence then they are said to have the ***same cardinality***. ◊

With these two definitions in place we can see that Theorem 11.1.3 is nothing less than the statement that the real numbers are not countably infinite. Since it is certainly not finite, then we say that the set of real numbers is uncountable and therefore "bigger" than the natural numbers!

To see this let us suppose first that each real number appears in the sequence $(s_n)_{n=1}^{\infty}$ exactly once. In that case the indexing of our sequence is really just a one-to-one correspondence between the elements of the sequence and $\mathbb{N}$:

$$\begin{aligned} 1 &\leftrightarrow s_1 \\ 2 &\leftrightarrow s_2 \\ 3 &\leftrightarrow s_3 \\ 4 &\leftrightarrow s_4 \\ &\vdots \end{aligned}$$

If some real numbers are repeated in our sequence then all of the real numbers are a subset of our sequence and will therefore also be countable (see Problem 11.1.7, part a).

In either case, every sequence is countable. But our theorem says that *no* sequence in $\mathbb{R}$ includes all of $\mathbb{R}$. Therefore $\mathbb{R}$ is uncountable.

Most of the sets you have encountered so far in your life have been countable.

Problem 11.1.6 Show that each of the following sets is countable.

(a) $\{2, 3, 4, 5, \ldots\} = \{n\}_{n=2}^{\infty}$

(b) $\{0, 1, 2, 3, \ldots\} = \{n\}_{n=0}^{\infty}$

(c) $\{1, 4, 9, 16, \ldots, n^2, \ldots\} = \{n^2\}_{n=1}^{\infty}$

(d) The set of prime numbers

(e) $\mathbb{Z}$

□

In fact, if we start with a countable set it is rather difficult to use it to build anything but another countable set.

Problem 11.1.7 Let $\{A_1, A_2, A_3, \ldots\}$ be a collection of countable sets. Show that each of the following sets is also countable:

(a) Any subset of A_1.

(b) $A_1 \cup A_2$

(c) $A_1 \cup A_2 \cup A_3$

(d) $\displaystyle\bigcup_{i=1}^{n} A_i$

(e) $\displaystyle\bigcup_{i=1}^{\infty} A_i$

□

It seems that no matter what we do the only example of an uncountably infinite set is $\mathbb{R}$. But wait! Remember the rational numbers? They were similar to the real numbers in many ways. Perhaps they are uncountably infinite too?

Alas, no. The rational numbers turn out to be countable too.

Theorem 11.1.8 *Show that* $\mathbb{Q}$ *is countable.*

Sketch of Proof. First explain how you know that all of the non-negative rational numbers are in this list:

$$\frac{0}{1}, \frac{0}{2}, \frac{1}{1}, \frac{0}{3}, \frac{1}{2}, \frac{2}{1}, \frac{0}{4}, \frac{1}{3}, \frac{2}{2}, \frac{3}{1}, \ldots.$$

However there is clearly some duplication. To handle this, apply part (a) of Problem 11.1.7. Does this complete the proof or is there more to do? ■

Problem 11.1.9 Prove Theorem 11.1.8. □

The following corollary says that the cardinality of the real numbers is *much* larger than the cardinality of the rational numbers, despite the fact that both are infinite.

That is, as a subset of the reals, the rationals can be contained in a sequence of intervals, the sum of whose lengths can be arbitrarily small. In a sense this says that a countably infinite set is so small (on the transfinite scale) that it is "almost" finite.

Usually we express this idea with the statement, "$\mathbb{Q}$ is a set of *measure zero* in $\mathbb{R}$". The term "measure" has a precise meaning which we will not pursue. The following corollary contains the essence of the idea.

Corollary 11.1.10 *Let $\varepsilon > 0$ be given. There is a collection of intervals in $\mathbb{R}$, $I_n = [a_n, b_n]$ such that*

$$\mathbb{Q} \subset \bigcup_{n=1}^{\infty} I_n$$

and

$$\sum_{n=1}^{\infty} (b_n - a_n) < \varepsilon.$$

Problem 11.1.11 Prove Corollary 11.1.10.

Hint.

If we had only finitely many rationals to deal with this would be easy. Let $\{r_1, r_2, \ldots, r_k\}$ be these rational numbers and take $a_n = r_n - \frac{\varepsilon}{2(k+1)}$ and $b_n = r_n + \frac{\varepsilon}{2(k+1)}$. Then for all $n = 1, \ldots, k$ $r_n \in [a_n, b_n]$ and

$$\sum_{n=1}^{k} b_n - a_n = \sum_{n=1}^{k} \frac{\varepsilon}{k+1} = \frac{k\varepsilon}{k+1} < \varepsilon.$$

The difficulty is, how do we move from the finite to the infinite case?] □

Notice how this idea hearkens back to the discussion of Leibniz's approach to the Product Rule (equation (I.1)). He simply tossed aside the expression $dx\,dy$ because it was *infinitely small* compared to either $x\,dy$ or $y\,dx$. Although this isn't quite the same thing we are discussing here it is similar and it is clear that Leibniz's insight and intuition were extremely acute. They were moving him in the right direction, at least.

All of our efforts to build an uncountable set from a countable one have come to nothing. In fact many sets that at first "feel" like they should be uncountable are in fact countable. This makes the uncountability of $\mathbb{R}$ all the more remarkable.

Comment. The failure is in the methods we've used so far. It is possible to build an uncountable set using just two symbols if we're clever enough, but this would take us too far away from our main topic.

However if we *start* with an uncountable set it is relatively easy to build others from it.

Problem 11.1.12

(a) Let (a, b) and (c, d) be two open intervals of real numbers. Show that these two sets have the same cardinality by constructing a one-to-one onto function between them.

Hint.

A linear function should do the trick.

(b) Show that any open interval of real numbers has the same cardinality as $\mathbb{R}$.

Hint.

Consider the interval $(-\pi/2, \pi/2)$.

(c) Show that $(0, 1]$ and $(0, 1)$ have the same cardinality.

Hint.

Note that $\{1, 1/2, 1/3, \ldots\}$ and $\{1/2, 1/3, \ldots\}$ have the same cardinality.

(d) Show that $[0, 1]$ and $(0, 1)$ have the same cardinality.

□

11.2 Cantor's Theorem and Its Consequences

Figure 11.2.1 Richard Dedekind[19]

Once Cantor showed that there were two types of infinity (countable and uncountable), the following question was natural, "Do all uncountable sets have the same cardinality?"

Just like not all "non-dogs" are cats, there is, offhand, no reason to believe that all uncountable sets should be the same size. However constructing uncountable sets of different sizes is not as easy as it sounds.

For example, what about the line segment represented by the interval $[0, 1]$ and the square represented by the set $[0, 1] \times [0, 1] = \{(x, y)|0 \leq x, y \leq 1\}$. Certainly the two dimensional square must be a larger infinite set than the one dimensional line segment. Remarkably, Cantor showed that these two sets were the same cardinality. In his 1877 correspondence of this result to his friend and fellow mathematician, Richard Dedekind, even Cantor remarked, "I see it, but I don't believe it!"

[19]mathshistory.st-andrews.ac.uk/Biographies/Dedekind/

The following gives the original idea of Cantor's proof. Cantor devised the following function $f : [0,1] \times [0,1] \to [0,1]$. First, we represent the coordinates of any point $(x, y) \in [0,1] \times [0,1]$ by their decimal representations $x = 0.a_1a_2a_3\ldots$ and $y = 0.b_1b_2b_3\ldots$. Even terminating decimals can be written this way as we could write $0.5 = 0.5000\ldots$. We can then define $f(x, y)$ by

$$f((0.a_1a_2a_3\ldots, 0.b_1b_2b_3\ldots)) = 0.a_1b_1a_2b_2a_3b_3\ldots.$$

This relatively simple idea has some technical difficulties in it related to the following result.

Problem 11.2.2 Consider the sequence $(0.9, 0.99, 0.999, \ldots)$. Determine that this sequence converges and, in fact, it converges to 1. This suggests that $0.999\ldots = 1$. □

Similarly, we have $0.04999\ldots = 0.05000\ldots$, etc. To make the decimal representation of a real number in $[0,1]$ unique, we must make a consistent choice of writing a terminating decimal as one that ends in an infinite string of zeros or an infinite string of nines [with the one exception $0 = 0.000\ldots$]. No matter which choice we make, we could never make this function onto. For example, $109/1100 = 0.09909090\ldots$ would have as its pre-image $(0.0999\ldots, 0.9000\ldots)$ which would be a mix of the two conventions.

Cantor was able to overcome this technicality to demonstrate a one to one correspondence, but instead we will note that in either convention, the function is one-to-one, so this says that the set $[0,1] \times [0,1]$ is the same cardinality as some (uncountable) subset of $\mathbb{R}$. The fact that this has the same cardinality as $\mathbb{R}$ is something we will come back to. But first we'll try construct an uncountable set which does not have the same cardinality as $\mathbb{R}$. To address this issue, Cantor proved the following in 1891.

Theorem 11.2.3 ***Cantor's Theorem***

Let S be any set. Then there is no one-to-one correspondence between S and $P(S)$, the set of all subsets of S.

Since S can be put into one-to-one correspondence with a subset of $P(S)$ $(a \to \{a\})$, then this says that $P(S)$ is at least as large as S. In the finite case $|P(S)|$ is strictly greater than $|S|$ as the following problem shows. It also demonstrates why $P(S)$ is called the power set of S.

Problem 11.2.4 Prove: If $|S| = n$, then $|P(S)| = 2^n$.

Hint.

Let $S = \{a_1, a_2, \ldots, a_n\}$. Consider the following correspondence between the elements of $P(S)$ and the set T of all n-tuples of yes (Y) or no (N):

$$\begin{aligned} \{\} &\leftrightarrow \{N, N, N, \ldots, N\} \\ \{a_1\} &\leftrightarrow \{Y, N, N, \ldots, N\} \\ \{a_2\} &\leftrightarrow \{N, Y, N, \ldots, N\} \\ &\vdots \\ S &\leftrightarrow \{Y, Y, Y, \ldots, Y\} \end{aligned}$$

How many elements are in T? □

Problem 11.2.5 Prove Cantor's Theorem 11.2.3.

Hint.

Assume for contradiction, that there is a one-to-one correspondence $f : S \to P(S)$. Consider $A = \{x \in S | x \notin f(x)\}$. Since f is onto, then there is $a \in A$ such that $A = f(a)$. Is $a \in A$ or is $a \notin A$? □

Actually it turns out that $\mathbb{R}$ and $P(\mathbb{N})$ have the same cardinality. This can be seen in a roundabout way using some of the above ideas from Problem 11.2.4. Specifically, let T be the set of all sequences of zeros or ones (you can use Ys or Ns, if you prefer). Then it is straightforward to see that T and $P(\mathbb{N})$ have the same cardinality.

If we consider $(0, 1]$, which has the same cardinality as $\mathbb{R}$, then we can see that this has the same cardinality as T as well. Specifically, if we think of the numbers in binary, then every real number in $[0, 1]$ can be written as $\sum_{j=1}^{\infty} \frac{a_j}{2^j} = 0.a_1a_2a_3\ldots$ where $a_j \in \{0, 1\}$. We have to account for the fact that binary representations such as $0.0111\ldots$ and $0.1000\ldots$ represent the same real number (say that no representations will end in an infinite string of zeros), then we can see that $[0, 1]$ has the same cardinality as $T - U$, where U is the set of all sequences ending in an infinite string of zeros. It turns out that U itself is a countable set.

Problem 11.2.6 Let $U_n = \{(a_1, a_2, a_3, \ldots) \mid a_j \in \{0, 1\} \text{ and } a_{n+1} = a_{n+2} = \cdots = 0\}$. Show that for each n, U_n is finite and use this to conclude that U is countably infinite. □

The above problems say that $\mathbb{R}$, $T - U$, T, and $P(N)$ all have the same cardinality. The following two problems show that deleting a countable set from an uncountable set does not change its cardinality.

Problem 11.2.7 Let S be an infinite set. Prove that S contains a countably infinite subset. □

Problem 11.2.8 Suppose X is an uncountable set and $Y \subset X$ is countably infinite.

(a) Prove that X and $X - Y$ are both uncountable sets.

(b) Prove that X and $X - Y$ have the same cardinality.

Hint.

Let $Y = Y_0$. If $X - Y_0$ is an infinite set, then by the previous problem it contains a countably infinite set Y_1. Likewise if $X - (Y_0 \cup Y_1)$ is infinite it also contains an infinite set Y_2. Again, if $X - (Y_0 \cup Y_1 \cup Y_2)$ is an infinite set then it contains an infinite set Y_3, etc. For $n = 1, 2, 3, \ldots$, let $f_n : Y_{n-1} \to Y_n$ be a one-to-one correspondence and define $f : X \to X - Y$ by

$$f(x) = \begin{cases} f_n(x), & \text{if } x \in Y_n, n = 0, 1, 2, \ldots \\ x, & \text{if } x \in X - (\cup_{n=0}^{\infty} Y_n) \end{cases}.$$

Show that f is one-to-one and onto.

□

As we indicated before, Cantor's work on infinite sets had a profound impact on mathematics in the beginning of the twentieth century. For example, in examining the proof of Cantor's Theorem, the eminent logician Bertrand Russell devised his famous paradox in 1901. Before this time, a set was naively thought of as just a collection of objects. Through the work of Cantor and others, sets were becoming a central object of study in mathematics as many mathematical concepts were being reformulated in terms of sets. The idea was that set theory was to be a unifying theme of mathematics. This paradox set the mathematical world on its ear.

Russell's Paradox: Consider the set of all sets which are not elements of themselves. We call this set D and ask, "Is $D \in D$?" Symbolically, this set is

$$D = \{S \mid S \notin S\}.$$

If $D \in D$, then by definition, $D \notin D$. If $D \notin D$, then by definition, $D \in D$.

If you look back at the proof of Cantor's Theorem, this was basically the idea that gave us the contradiction. To have such a contradiction occurring at the most basic level of mathematics was scandalous. It forced a number of mathematicians and logicians to carefully devise the axioms by which sets could be constructed. To be honest, most mathematicians still approach set theory from a naive point of view as the sets we typically deal with fall under the category of what we would call "normal sets." In fact, such an approach is officially called Naive Set Theory (as opposed to Axiomatic Set Theory). However, attempts to put set theory and logic on solid footing led to the modern study of symbolic logic and ultimately the design of computer (machine) logic.

Another place where Cantor's work had a profound influence in modern logic comes from something we alluded to before. We showed before that the unit square $[0, 1] \times [0, 1]$ had the same cardinality as an uncountable subset of $\mathbb{R}$. In fact, Cantor showed that the unit square had the same cardinality as $\mathbb{R}$ itself and was moved to advance the following in 1878.

Conjecture 11.2.9 The Continuum Hypothesis. *Every uncountable subset of $\mathbb{R}$ has the same cardinality as $\mathbb{R}$.*

Cantor was unable to prove or disprove this conjecture (along with every other mathematician). In fact, proving or disproving the Continuum Hypothesis, was one of Hilbert's famous 23 problems presented as a challenge to mathematicians at the International Congress of Mathematicians in 1900.

Since $\mathbb{R}$ has the same cardinality as $P(\mathbb{N})$, then the Continuum Hypothesis was generalized to the:

Conjecture 11.2.10 The Generalized Continuum Hypothesis. *Given an infinite set S, there is no infinite set which has a cardinality strictly between that of S and its power set $P(S)$.*

The Zermelo-Fraenkel Axioms. One of the formal axiomatic approaches to set theory established by Ernst Zermelo in 1908 and revised by Abraham Fraenkel in 1921.

Efforts to prove or disprove this were in vain and with good reason. In 1940, the logician Kurt Gödel showed that the Continuum Hypothesis could not be disproved from the Zermelo-Fraenkel Axioms of set theory. In 1963, Paul Cohen showed that the Continuum Hypothesis could not be proved using the Zermelo-Fraenkel Axioms. In other words, the Zermelo-Fraenkel Axioms do not contain enough information to decide the truth of the hypothesis.

We are willing to bet that at this point your head might be swimming a bit with uncertainty. If so, then know that these are the same feelings that the mathematical community experienced in the mid twentieth century. In the past, mathematics was seen as a model of logical certainty. It is disconcerting to find that there are statements that are "undecidable." In fact, Gödel proved in 1931 that a consistent finite axiom system that contained the axioms of arithmetic would always contain undecidable statements which could neither be proved true nor false with those axioms. Mathematical knowledge would always be incomplete.

Figure 11.2.11 Kurt Gödel[20]

So by trying to put the foundations of calculus on solid ground, we have come to a point where we can never obtain mathematical certainty. Does this mean that we should throw up our hands and concede defeat? Should we be paralyzed with fear of trying anything? Certainly not! As we mentioned before, most mathematicians do well by taking a pragmatic approach: using their mathematics to solve problems that they encounter. In fact, it is typically the problems that motivate the mathematics. It is true that mathematicians take chances that don't always pan out, but they still take these chances, often with success. Even when the successes lead to more questions, as they typically do, tackling those questions usually leads to a deeper understanding. At the very least, our incomplete understanding means we will always have more questions to answer, more problems to solve.

What else could a mathematician ask for?

[20] mathshistory.st-andrews.ac.uk/Biographies/Godel/

Chapter 12

Epilogues

12.1 On the Nature of Numbers: A Dialogue (with Apologies to Galileo)

Interlocuters:*Salviati, Sagredo, and Simplicio; Three Friends of Galileo Galilei*

Setting: Three friends meet in a garden for lunch in Renassaince Italy. Prior to their meal they discuss the book *How We Got From There to Here: A Story of Real Analysis.* How they obtained a copy is not clear.

Salviati: My good sirs. I have read this very strange volume as I hope you have?

Sagredo: I have and I also found it very strange.

Simplicio: Very strange indeed; at once silly and mystifying.

Salviati: Silly? How so?

Simplicio: These authors begin their tome with the question, "What is a number?" This is an unusually silly question, don't you think? Numbers are numbers. Everyone knows what they are.

Sagredo: I thought so as well until I reached the last chapter. But now I am not so certain. What about this quantity $\aleph_0$? If this counts the positive integers, isn't it a number? If not, then how can it count anything? If so, then what number is it? These questions plague me 'til I scarcely believe I know anything anymore.

Simplicio: Of course $\aleph_0$ is not a number! It is simply a new name for the infinite, and infinity is not a number.

Sagredo: But isn't $\aleph_0$ the cardinality of the set of natural numbers, $\mathbb{N}$, in just the same way that the cardinality of the set $S = \{Salviati, Sagredo, Simplicio\}$ is 3? If 3 is a number, then why isn't $\aleph_0$?

Simplicio: Ah, my friend, like our authors you are simply playing with words. You count the elements in the set $S = \{Salviati, Sagredo, Simplicio\}$; you see plainly that the *number of elements* it contains is 3 and then you change your language. Rather than saying that the *number of elements* in S is 3 you say that the *cardinality* is 3. But clearly "cardinality" and "number of elements" mean the same thing.

Similarly you use the symbol $\mathbb{N}$ to denote the set of positive integers. With your new word and symbol you make the statement "the *cardinality* (number of elements) of $\mathbb{N}$ is $\aleph_0$." This statement has the same grammatical form as the statement "the *number of elements* (cardinality) of S is three." Since three *is* a number you conclude that $\aleph_0$ is also a number.

But this is simply nonsense dressed up to sound sensible. If we unwind our notation and language, your statement is simply, "The number of positive integers is infinite." This is obviously nonsense because infinity is *not* a number.

Even if we take infinity as an undefined term and try to define it by your statement this is still nonsense since you are using the word "number" to define a new "number" called infinity. This definition is circular. Thus it is no definition at all. It is nonsense.

Salviati: Your reasoning on this certainly seems sound.

Simplicio: Thank you.

Salviati: However, there are a couple of small points I would like to examine more closely if you will indulge me?

Simplicio: Of course. What troubles you?

Salviati: You've said that we cannot use the word "number" to define numbers because this would be circular reasoning. I entirely agree, but I am not sure this is what our authors are doing.

Consider the set $\{1, 2, 3\}$. Do you agree that it contains three elements?

Simplicio: Obviously.

Sagredo: Ah! I see your point! That there are three elements does *not* depend on what those elements are. Any set with three elements has three elements regardless of the nature of the elements. Thus saying that the set $\{1, 2, 3\}$ contains three elements does not define the word "number" in a circular manner because it is irrelevant that the *number* 3 is one of the elements of the set. Thus to say that three is the cardinality of the set $\{1, 2, 3\}$ has the same meaning as saying that there are three elements in the set $\{Salviati, Sagredo, Simplicio\}$. In both cases the number "3" is the name that we give to the totality of the elements of each set.

Salviati: Precisely. In exactly the same way $\aleph_0$ is the symbol we use to denote the totality of the set of positive integers.

Thus $\aleph_0$ is a number in the same sense that '3' is a number, is it not?

Simplicio: I see that we can say in a meaningful way that three is the cardinality of any set with . . . well, . . . with three elements (it becomes very difficult to talk about these things) but this is simply a tautology! It is a way of saying that a set which has three elements has three elements!

This means only that we have counted them and we had to stop at three. In order to do this we must have numbers first. Which, of course, we do. As I said, everyone knows what numbers are.

Sagredo: I must confess, my friend, that I become more confused as we speak. I am no longer certain that I really know what a number is. Since you seem to have retained your certainty can you clear this up for me? Can you tell me what a number is?

Simplicio: Certainly. A number is what we have just been discussing. It is what you have when you stop counting. For example, three is the totality (to use your phrase) of the elements of the sets $\{Salviati, Sagredo, Simplicio\}$ or $\{1, 2, 3\}$ because when I count the elements in either set I have to stop at three. Nothing less, nothing more. Thus three is a number.

Salviati: But this definition only confuses me! Surely you will allow that fractions are numbers? What is counted when we end with, say $4/5$ or $1/5$?

Simplicio: This is simplicity itself. $4/5$ is the number we get when we have divided something into 5 equal pieces and we have counted four of these fifths. This is four-fifths. You see? Even the language we use naturally bends itself to our purpose.

Salviati: But what of one-fifth? In order to count one fifth we must first divide something into fifths. To do this we must know what one-fifth is, musn't we? We seem to be using the word "number" to define itself again. Have we not come full circle and gotten nowhere?

Simplicio: I confess this had not occurred to me before. But your objection is easily answered. To count one-fifth we simply divide our "something" into tenths. Then we count two of them. Since two-tenths is the same as one-fifth the problem is solved. Do you see?

Sagredo: I see your point but it will not suffice at all! It merely replaces the question, "What is one-fifth?" with, "What is one-tenth?" Nor will it do to say that one-tenth is merely two-twentieths. This simply shifts the question back another level.

Archimedes said, "Give me a place to stand and a lever long enough and I will move the earth." But of course he never moved the earth because he had nowhere to stand. We seem to find ourselves in Archimedes' predicament: We have no place to stand.

Simplicio: I confess I don't see a way to answer this right now. However I'm sure an answer can be found if we only think hard enough. In the meantime I cannot accept that $\aleph_0$ is a number. It is, as I said before, infinity and infinity is not a number! We may as well believe in fairies and leprechauns if we call infinity a number.

Sagredo: But again we've come full circle. We cannot say definitively that $\aleph_0$ is or is not a number until we can state with confidence what a number is. And even if we could find solid ground on which to solve the problem of fractions, what of $\sqrt{2}$? Or π? Certainly these are numbers but I see no way to count to either of them.

Simplicio: Alas! I am beset by demons! I am bewitched! I no longer believe what I *know* to be true!

Salviati: Perhaps things are not quite as bad as that. Let us consider further. You said earlier that we all know what numbers are, and I agree. But perhaps your statement needs to be more precisely formulated. Suppose we say instead that we all know what numbers *need* to be? Or that we know what we *want* numbers to be?

Even if we cannot say with certainly what numbers *are* surely we can say what we want and need for them to be. Do you agree?

Sagredo: I do.

Simplicio: And so do I.

Salviati: Then let us invent numbers anew, as if we've never seen them before, always keeping in mind those properties we need for numbers to have. If we take this as a starting point then the question we need to address is, "What do we need numbers to be?"

Sagredo: This is obvious! We need to be able to add them and we need to be able to multiply them together, and the result should also be a number.

Simplicio: And subtract and divide too, of course.

Sagredo: I am not so sure we actually need these. Could we not *define* "subtract two from three" to be "add negative two to three" and thus dispense with subtraction and division?

Simplicio: I suppose we can but I see no advantage in doing so. Why not simply have subtraction and division as we've always known them?

Sagredo: The advantage is parsimony. Two arithmetic operations are easier to keep track of than four. I suggest we go forward with only addition and multiplication for now. If we find we need subtraction or division we can consider them later.

Simplicio: Agreed. And I now see another advantage. Obviously addition and multiplication must not depend on order. That is, if x and y are numbers then $x+y$ must be equal to $y+x$ and xy must be equal to yx. This is not true for subtraction, for $3-2$ does *not* equal $2-3$. But if we define subtraction as you suggest then this symmetry is preserved:

$$x + (-y) = (-y) + x.$$

Sagredo: Excellent! Another property we will require of numbers occurs to me now. When adding or multiplying more than two numbers it should not matter where we begin. That is, if x, y and z are numbers it should be true that

$$(x + y) + z = x + (y + z)$$

and

$$(x \cdot y) \cdot z = x \cdot (y \cdot z).$$

Simplicio: Yes! We have it! Any objects which combine in these precise ways can be called numbers.

Salviati: Certainly these properties are necessary, but I don't think they are yet sufficient to our purpose. For example, the number 1 is unique in that it is the only number which, when multiplying another number leaves it unchanged.

For example: $1 \cdot 3 = 3$. Or, in general, if x is a number then $1 \cdot x = x$.

Sagredo: Yes. Indeed. It occurs to me that the number zero plays a similar role for addition: $0 + x = x$.

Salviati: It does not seem to me that addition and multiplication, as we have defined them, force 1 or 0 into existence so I believe we will have to postulate their existence independently.

Sagredo: Is this everything then? Is this all we require of numbers?

Simplicio: I don't think we are quite done yet. How shall we get division?

Sagredo: In the same way that we defined subtraction to be the addition of a negative number, can we not define division to be multiplication by a reciprocal? For example, 3 divided by 2 can be considered 3 *multiplied* by 1/2, can it not?

Salviati: I think it can. But observe that *every* number will need to have a corresponding negative so that we can subtract any amount. And again nothing we've discussed so far forces these negative numbers into existence so we will have to postulate their existence separately.

Simplicio: And in the same way every number will need a reciprocal so that we can divide by any amount.

Sagredo: Every number that is, except zero.

Simplicio: Yes, this is true. Strange is it not, that of them all only this one number needs no reciprocal? Shall we also postulate that zero has no reciprocal?

Salviati: I don't see why we should. Possibly $\aleph_0$ is the reciprocal of zero. Or possibly not. But I see no need to concern ourselves with things we do not need.

Simplicio: Is this everything then? Have we discovered all that we need for numbers to be?

Salviati: I believe there is only one property missing. We have postulated addition and we have postulated multiplication and we have described the numbers zero and one which play similar roles for addition and multiplication respectively. But we have not described how addition and multiplication work together.

That is, we need a rule of distribution: If x, y and z are all numbers then

$$x \cdot (y + z) = x \cdot y + x \cdot z.$$

With this in place I believe we have everything we need.

Simplicio: Indeed. We can also see from this that $\aleph_0$ cannot be a number since, in the first place, it cannot be added to another number and in the second, even if it could be added to a number the result is surely not also a number.

Salviati: My dear Simplicio, I fear you have missed the point entirely! Our axioms do not declare what a number is, only how it behaves with respect to addition and multiplication with other numbers. Thus it is a mistake to presume that "numbers" are only those objects that we have always believed them to be. In fact, it now occurs to me that "addition" and "multiplication" also needn't be seen as the operations we have always believed them to be.

For example suppose we have three objects, $\{a, b, c\}$ and suppose that we define "addition" and "multiplication" by the following tables:

$+$	a	b	c
a	a	b	c
b	b	c	a
c	c	a	b

$\cdot$	a	b	c
a	a	a	a
b	a	b	c
c	a	c	b

I submit that our set along with these definitions satisfy all of our axioms and thus a, b and c qualify to be called "numbers."

Simplicio: This cannot be! There is no zero, no one!

Sagredo: But there is. Do you not see that a plays the role of zero — if you add it to any number you get that number back. Similarly b plays the role of one.

Simplicio: This is astonishing! If a, b and c can be numbers then I am less sure than ever that I know what numbers are! Why, if we replace a, b, and c with *Simplicio Sagredo,* and *Salviati* then *we* become numbers ourselves!

Salviati: Perhaps we will have to be content with knowing how numbers behave rather than knowing what they are.

However I confess that I have a certain affection for the numbers I grew up with. Let us call those the "real" numbers. Any other set of numbers, such as our $\{a, b, c\}$ above we will call a field of numbers, since they seem to provide us with new ground to explore. Or perhaps just a *number field*?

As we have been discussing this I have been writing down our axioms. They are these. Numbers are any objects which satisfy all of the following properties:

AXIOMS OF NUMBERS

Axiom I: Definition of Operations They can be combined by two operations, denoted "$\cdot$" and "+".

Axiom II: Closure If x, y and z are numbers then $x + y$ is also a number. $x \cdot y$ is also a number.

Axiom II: Commutativity $x + y = y + x$ $x \cdot y = y \cdot x$

Axiom III: Associativity $(x + y) + z = x + (y + z)$ $(x \cdot y) \cdot z = x \cdot (y \cdot z)$

Axiom IV: Additive Identity There is a number, denoted 0, such that for any number, x,

$$x + 0 = x.$$

Axiom V: Multiplicative Identity There is a number, denoted 1, such that for any number, x,

$$1 \cdot x = x.$$

Axiom VI: Additive Inverses Given any number, x, there is a number, denoted $-x$, with the property that

$$x + (-x) = 0.$$

Axiom VII: Multiplicative Inverse Given any number, $x \neq 0$, there is a number, denoted x^{-1}, with the property that

$$x \cdot x^{-1} = 1.$$

Axiom VIII: The Distributive Property If x, y and z are numbers then

$$x \cdot (y + z) = x \cdot y + x \cdot z.$$

Sagredo: My friend, this is a thing of surpassing beauty! All seems clear to me now. Numbers are any group of objects which satisfy our axioms. That is, a number is anything that *acts* like a number.

Salviati: Yes this seems to be true.

Simplicio: But wait! We have not settled the question: Is $\aleph_0$ a number or not?

Salviati: If everything we have just done is valid then $\aleph_0$ *could be* a number. And so could $\aleph_1$,$\aleph_2 \ldots$ *if* we can find a way to define addition and multiplication on the set $\{\aleph_0, \aleph_1, \aleph_2 \ldots\}$ in a manner that agrees with our axioms.

Sagredo: An arithmetic of infinities! This is a very strange idea. Can such a thing be made sensible?

Simplicio: Not, I think, before lunch. Shall we retire to our meal?

12.1.1 Additional Problems

Problem 12.1.1 Show that $0 \neq 1$.

Hint.

Show that if $x \neq 0$, then $0 \cdot x \neq x$. □

Problem 12.1.2 Consider the set of ordered pairs of integers: $\{(x, y)|x, y \in \mathbb{Z}\}$, and define addition and multiplication as follows:

- *Addition:.*

  $(a, b) + (c, d) = (ad + bc, bd)$

- *Multiplication:.*

  $(a, b) \cdot (c, d) = (ac, bd)$.

(a) If we add the convention that

$$(ab, ad) = (b, d)$$

show that this set with these operations forms a number field.

(b) Which number field is this?

□

Problem 12.1.3 Consider the set of ordered pairs of real numbers, $\{(x, y)|x, y \in \mathbb{R}\}$, and define addition and multiplication as follows:

- *Addition:.*

  $(a, b) + (c, d) = (a + c, b + d)$

- *Multiplication:.*
  $(a, b) \cdot (c, d) = (ac - bd, ad + bc)$.

(a) Show that this set with these operations forms a number field.

(b) Which number field is this?

□

12.2 Building the Real Numbers

Contrary to the title of this section we will not be rigorously building the real numbers here. Instead our goal is to show why such a build is logically necessary, and to give a sense of some of the ways this has been accomplished in the past. This may seem odd given our uniform emphasis on mathematical rigor, especially in the third part of the text, but there are very good reasons for this.

One is simple practicality. The fact is that rigorously building the real numbers and then showing that they have the required properties is extraordinarily detailed work, even for mathematics. If we want to keep this text to a manageable size (we do), we simply don't have the room.

The second reason is that there is, as far as we know, very little for you to gain by it. When we are done we will have the real numbers. The same real numbers you have been using all of your life. They have the same properties, and quirks, they've always had. To be sure, they will not have lost any of their charm; They will be the same delightful mix of the mundane and the bizarre, and they are still well worth exploring and getting to know better. But nothing we do in the course of building them up logically from simpler ideas will help with that exploration.

A reasonable question then, is, "Why bother?" If the process is overwhelmingly, tediously detailed (it is) and gives us nothing new for our efforts, why do it at all?

Andrew Wiles. The man who proved Fermat's Last Theorem[21].

Andrew Wiles[22] has compared doing mathematics has been compared to entering a dark room. At first you are lost. The layout of the room and furniture are unknown so you fumble about for a bit and slowly get a sense of your immediate environs, perhaps a vague notion of the organization of the room as a whole. Eventually, after much, often tedious exploration, you become quite comfortable in your room. But always there will be dark corners; hidden areas you have not yet explored. Such a dark area may hide anything; the latches to unopened doors you didn't know were there; a clamp whose presence explains why you couldn't move that little desk in the corner; even the light switch that would allow you to illuminate an area more clearly than you would have imagined possible.

But, and this is the point, there is no way to know what you will find there until you walk into that dark corner and begin exploring. Perhaps nothing. But perhaps something wonderful.

This is what happened in the late nineteenth century. The real numbers had been used since the Pythagoreans learned that $\sqrt{2}$ was irrational. But really, most calculuations were (and still are) done with just the rational numbers. Moreover, since $\mathbb{Q}$ forms a "set of measure zero," it is clear that most of the

[21] mathshistory.st-andrews.ac.uk/HistTopics/Fermat's_last_theorem/
[22] mathshistory.st-andrews.ac.uk/Biographies/Wiles/

real numbers had gone completely unused. The set of real numbers was thus one of those "dark corners" of mathematics. It had to be explored.

Measure Zero Sets. See Corollary 11.1.10 of Chapter 11.

"But even if that is true," you might ask, ``I have no interest in the logical foundations of the real numbers, especially if such knowledge won't tell me anything I don't already know. Why do I need to know all of the details of constructing $\mathbb{R}$ from $\mathbb{Q}$?

The answer to this is very simple: You don't.

That's the other reason we're not covering all of the details of this material. We will explain enough to light up, dimly perhaps, this little corner of mathematics. Later, should you need (or want) to come back to this and explore further you will have a foundation to start with. Nothing more.

Until the nineteenth century the geometry of Euclid, as given in his book *The Elements,* was universally regarded as the touchstone of mathematical perfection. This belief was so deeply embedded in Western culture that as recently as 1923, Edna St. Vincent Millay opened one of the poems in her book *The Harp Weaver and Other Poems* with the line "Euclid alone has looked on beauty bare."

Euclid begins his book by stating 5 simple axioms and proceeds, step by logical step, to build up his geometry. Although far from actual perfection, his methods are clean, precise and efficient — he arrives at the Pythagorean Theorem in only 47 steps (theorems) — and even today Euclid's *Elements* still sets a very high standard of mathematical exposition and parsimony.

The goal of starting with what is clear and simple and proceeding logically, rigorously, to what is complex is still a guiding principle of all mathematics for a variety of reasons. In the late nineteenth century, this principle was brought to bear on the real numbers. That is, some properties of the real numbers that at first seem simple and intuitively clear turn out on closer examination, as we have seen, to be rather counter-intuitive. This alone is not really a problem. We can have counter-intuitive properties in our mathematics — indeed, this is a big part of what makes mathematics interesting — as long as we arrive at them logically, starting from simple assumptions the same way Euclid did.

Having arrived at a view of the real numbers which is comparable to that of our nineteenth century colleagues, it should now be clear that the real numbers and their properties must be built up from simpler concepts as suggested by our Italian friends in the previous section.

In addition to those properties we have discovered so far, both $\mathbb{Q}$ and $\mathbb{R}$ share another property which will be useful. We have used it throughout this text but have not heretofore made it explicit. They are both *linearly ordered.* We will now make this property explicit.

Definition 12.2.1 Linear Ordering. A number field is said to be *linearly ordered* if there is a relation, denoted "$<$", on the elements of the field which satisfies all of the following for all x, y, and z in the field.

1. For all numbers x and y in the field, exactly one of the following holds:

   (a) $x < y$

   (b) $x = y$

   (c) $y < x$

2. If $x < y$, then $x + z < y + z$ for all z in the field.

3. If $x < y$, and $0 < z$, then $x \cdot z < y \cdot z$.

4. If $x < y$ and $y < z$ then $x < z$.

$\diamond$

Any number field with such a relation is called a **linearly ordered number field** and as the following problem shows, not every number field is linearly ordered.

Problem 12.2.2

(a) Prove that the following must hold in *any* linearly ordered number field.

(a) $0 < x$ if and only if $-x < 0$.

(b) If $x < y$ and $z < 0$ then $y \cdot z < x \cdot z$.

(c) For all $x \neq 0$, $0 < x^2$.

(d) $0 < 1$.

(b) Show that the set of complex numbers ($\mathbb{C}$) is not a linearly ordered field.

$\square$

In a thorough, rigorous presentation we would now assume the existence of the natural numbers ($\mathbb{N}$), and their properties and use these to define the integers, ($\mathbb{Z}$). We would then use the integers to define the rational numbers, ($\mathbb{Q}$). We could then show that the rationals satisfy the field axioms worked out in the previous section, and that they are linearly ordered.

Then — at last — we would use $\mathbb{Q}$ to define the real numbers ($\mathbb{R}$), show that these also satisfy the field axioms and also have the other properties we expect: Continuity, the Nested Interval Property, the Least Upper Bound Property, the Bolzano-Weierstrass Theorem, the convergence of all Cauchy sequences, and linear ordering.

We would start with the natural numbers because they seem to be simple enough that we can simply assume their properties. As Leopold Kronecker (1823-1891) said: "God made the natural numbers, all else is the work of man."

Unfortunately this is rather a lot to fit into this epilogue so we will have to abbreviate the process rather severely.

We will assume the existence and properties of the rational numbers. Building $\mathbb{Q}$ from the integers is not especially hard and it is easy to show that they satisfy the axioms worked out by Salviati, Sagredo and Simplicio in the previous section. But the level of detail required for rigor quickly becomes onerous.

Even starting at this fairly advanced position in the chain of logic there is still a considerable level of detail needed to complete the process. Therefore our exposition will necessarily be incomplete.

Rather than display, in full rigor, how the real numbers can be built up from the rationals we will show, in fairly broad terms, three ways this has been done in the past. We will give references later in case you'd like to follow up and learn more.

12.2.1 The Decimal Expansion

This is by far the most straightforward method we will examine. Since we begin with $\mathbb{Q}$, we already have some numbers whose decimal expansion is infinite. For example, $\frac{1}{3} = 0.333\ldots$. We also know that if $x \in \mathbb{Q}$ then expressing x as a decimal gives either a finite or a *repeating infinite* decimal.

More simply, we can say that $\mathbb{Q}$ consists of the set of all decimal expressions which eventually repeat. (If it eventually repeats zeros then it is what we've called a finite decimal.)

We then define the real numbers to be the set of all infinite decimals, repeating or not.

It may feel as if all we have to do is define addition and multiplication in the obvious fashion and we are finished. This set with these definitions obviously satisfy all of the field axioms worked out by our Italian friends in the previous section. Moreover it seems clear that all of our equivalent completeness axioms are satisfied.

However, things are not quite as clear cut as they seem.

The primary difficulty in this approach is that the decimal representation of the real numbers is so familiar that everything we need to show *seems* obvious. But stop and think for a moment. Is it really obvious how to define addition and multiplication of *infinite* decimals? Consider the addition algorithm we were all taught in grade school. That algorithm requires that we line up two numbers at their decimal points:

$$\begin{array}{r} d_1d_2.d_3d_4 \\ +\ \delta_1\delta_2.\delta_3\delta_4. \end{array}$$

We then begin adding in the rightmost column and proceed to the left. *But if our decimals are infinite we can't get started because there is no rightmost column!*

A similar problem occurs with multiplication.

So our first problem is to define addition and multiplication in $\mathbb{R}$ in a manner that re-captures addition and multiplication in $\mathbb{Q}$.

This is not a trivial task.

One way to proceed is to recognize that the decimal notation we've used all of our lives is really shorthand for the sum of an infinite series. That is, if $x = 0.d_1d_2d_3\ldots$ where $0 \leq d_i \leq 9$ for all $i \in \mathbb{N}$ then

$$x = \sum_{i=1}^{\infty} \frac{d_i}{10^i}.$$

Addition is now apparently easy to define: If $x = \sum_{i=1}^{\infty} \frac{d_i}{10^i}$ and $y = \sum_{i=1}^{\infty} \frac{\delta_i}{10^i}$ then

$$x + y = \sum_{i=1}^{\infty} \frac{e_i}{10^i} \text{ where } e_i = d_i + \delta_i.$$

But there is a problem. Suppose for some $j \in \mathbb{N}$, $e_j = d_j + \delta_j > 10$. In that case our sum does not satisfy the condition $0 \leq e_j \leq 9$ so it is not even clear that the expression $\sum_{j=1}^{\infty} \frac{e_j}{10^j}$ represents a real number. That is, we may not have the closure property of a number field. We will have to define some sort of "carrying" operation to handle this.

Problem 12.2.3 Define addition on infinite decimals in a manner that is closed.

Hint.

Find an appropriate "carry" operation for our definition. □

A similar difficulty arises when we try to define multiplication. Once we have a notion of carrying in place, we could define multiplication as just the multiplication of series. Specifically, we could define

$$(0.a_1a_2a_3\ldots) \cdot (0.b_1b_2b_3\ldots) = \left(\frac{a_1}{10} + \frac{a_2}{10^2} + \cdots\right) \cdot \left(\frac{b_1}{10} + \frac{b_2}{10^2} + \cdots\right)$$

$$= \frac{a_1b_1}{10^2} + \frac{a_1b_2 + a_2b_1}{10^3} + \frac{a_1b_3 + a_2b_2 + a_3b_1}{10^4} + \cdots.$$

We could then convert this to a "proper" decimal using our carrying operation.

Again the devil is in the details to show that such algebraic operations satisfy everything we want them to. Even then, we need to worry about linearly ordering these numbers and our completeness axiom.

Another way of looking at this is to think of an infinite decimal representation as a (Cauchy) sequence of finite decimal approximations. Since we know how to add and multiply finite decimal representations, we can just add and multiply the individual terms in the sequences. Of course, there is no reason to restrict ourselves to only these specific types of Cauchy sequences, as we see in our next approach.

12.2.2 Cauchy Sequences

As we've seen, Georg Cantor began his career studying Fourier series and quickly moved on to more foundational matters in the theory of infinite sets.

But he did not lose his fascination with real analysis when he moved on. Like many mathematicians of his time, he realized the need to build $\mathbb{R}$ from $\mathbb{Q}$. He and his friend and mentor Richard Dedekind (who's approach we will see in the next section) both found different ways to build $\mathbb{R}$ from $\mathbb{Q}$.

Cantor started with Cauchy sequences in $\mathbb{Q}$.

That is, we consider the set of all Cauchy sequences of rational numbers. We would like to *define* each such sequence to be a real number. The goal should be clear. If $(s_n)_{n=1}^{\infty}$ is a sequence in $\mathbb{Q}$ which converges to $\sqrt{2}$ then we will call (s_n) the real number $\sqrt{2}$.

This probably seems a bit startling at first. There are a *lot* of numbers in (s_n) (countably infinitely many, to be precise) and we are proposing putting all of them into a big bag, tying it up in a ribbon, and calling the whole thing $\sqrt{2}$. It seems a very odd thing to propose, but recall from the discussion in the previous section that we left the concept of "number" undefined. Thus if we can take *any* set of objects and define addition and multiplication in such a way that the field axioms are satisfied, then those objects are legitimately numbers. To show that they are, in fact, the *real* numbers we will also need the completeness property.

A bag full of rational numbers works as well as anything *if* we can define addition and multiplication appropriately.

Our immediate problem though is not addition or multiplication but uniqueness. If we take one sequence (s_n) which converges to $\sqrt{2}$ and define it to be $\sqrt{2}$, what will we do with all of the other sequences that converge to $\sqrt{2}$?

Also, we have to be careful not to refer to any real numbers, like the square root of two for example, as we define the real numbers. This would be a circular — and thus useless — definition. Obviously though, we can refer to *rational* numbers, since these are the tools we'll be using.

The solution is clear. We take *all* sequences of rational numbers that converge to $\sqrt{2}$, throw them into our bag and call *that* $\sqrt{2}$. Our bag is getting pretty full now.

But we need to do this without using $\sqrt{2}$ because it is a real number. The following two definitions satisfy all of our needs.

Definition 12.2.4 Let $x = (s_n)_{n=1}^{\infty}$ and $y = (\sigma_n)_{n=1}^{\infty}$ be Cauchy sequences in $\mathbb{Q}$. x and y are said to be equivalent if they satisfy the following property: For every $\varepsilon > 0$, $\varepsilon \in \mathbb{Q}$, there is a rational number N such that for all $n > N$,

$n \in \mathbb{N}$,

$$|s_n - \sigma_n| < \varepsilon.$$

We will denote equivalence by writing, $x \equiv y$. ♢

Problem 12.2.5 Show that:

(a) $x \equiv x$

(b) $x \equiv y \Rightarrow y \equiv x$

(c) $x \equiv y$ and $y \equiv z \Rightarrow x \equiv z$

□

Definition 12.2.6 Every set of all equivalent Cauchy sequences defines a real number. ♢

A very nice feature of Cantor's method is that it is very clear how addition and multiplication should be defined.

Definition 12.2.7 If

$$x = \{ (s_n)_{k=1}^{\infty} | \, (s_n)_{k=1}^{\infty} \text{ is Cauchy in } \mathbb{Q} \}$$

and

$$y = \{ (\sigma_n)_{k=1}^{\infty} | \, (\sigma_n)_{k=1}^{\infty} \text{ is Cauchy in } \mathbb{Q} \}$$

then we define the following:

- *Addition:.*
  $x + y = \{ (t_n)_{k=1}^{\infty} | \, t_k = s_k + \sigma_k, \forall (s_n) \in x, \text{ and } (\sigma_n) \in y\}$
- *Multiplication:.*
  $x \cdot y = \{ (t_n)_{k=1}^{\infty} | \, t_k = s_k \sigma_k, \forall (s_n) \in x, \text{ and } (\sigma_n) \in y\}$

♢

The notation used in Definition 12.2.6 can be difficult to read at first, but basically it says that addition and multiplication are done component-wise. However since x and y consist of all equivalent sequences we have to take *every* possible choice of $(s_n) \in x$ and $(\sigma_n) \in y$, form the sum (product) $(s_n + \sigma_n)_{n=1}^{\infty}$ $((s_n \sigma_n)_{n=1}^{\infty})$ and then show that all such sums (products) are equivalent. Otherwise addition (multiplication) is not well-defined: It would depend on which sequence we choose to represent x and y.

Problem 12.2.8 Let x and y be real numbers (that is, let them be sets of equivalent Cauchy sequences in $\mathbb{Q}$). If (s_n) and (t_n) are in x and (σ_n) and (τ_n) are in y then

$$(s_n + \tau_n)_{n=1}^{\infty} \equiv (\sigma_n + t_n)_{n=1}^{\infty}.$$

□

Theorem 12.2.9 *Let* 0^* *be the set of Cauchy sequences in* $\mathbb{Q}$ *which are all equivalent to the sequence* $(0, 0, 0, \ldots)$*. Then*

$$0^* + x = x.$$

Proof. From Problem 12.2.8 it is clear that in forming $0^* + x$ we can choose any sequence in 0^* to represent 0^* and any sequence in x to represent x. (This is because any other choice will yield a sequence equivalent to $0^* + x$.)
Thus we choose $(0, 0, 0, \ldots)$ to represent 0^* and any element of x, say

$(x_1, x_2, \ldots)$, to represent x. Then

$$\begin{aligned}(0,0,0,\ldots) + (x_1, x_2, x_3, \ldots) &= (x_1, x_2, x_3, \ldots)\\ &= x.\end{aligned}$$

Since any other sequences taken from 0^* and x respectively, will yield a sum equivalent to x (see Problem 12.2.5) we conclude that

$$0^* + x = x.$$

■

Problem 12.2.10 Identify the set of equivalent Cauchy sequences, 1^*, such that

$$1^* \cdot x = x.$$

□

Problem 12.2.11 Let x, y, and z be real numbers (equivalent sets of Cauchy sequences). Show that with addition and multiplication defined as above we have:

(a) $x + y = y + x$

(b) $(x + y) + z = x + (y + z)$

(c) $x \cdot y = y \cdot x$

(d) $(x \cdot y) \cdot z = x \cdot (y \cdot z)$

(e) $x \cdot (y + z) = x \cdot y + x \cdot z$

□

Comment. We will not address this issue here, but you should give some thought to how this might be accomplished.

Once the existence of additive and multiplicative inverses is established the collection of all sets of equivalent Cauchy sequences, with addition and multiplication defined as above satisfy all of the field axioms. It is clear that they form a number field and thus deserve to be called numbers.

However this does not necessarily show that they form $\mathbb{R}$. We also need to show that they are complete in the sense of Chapter 9. It is perhaps not too surprising that when we build the real numbers using equivalent Cauchy sequences the most natural completeness property we can show is that if a sequence of real numbers is Cauchy then it converges.

However we are not in a position to show that Cauchy sequences in $\mathbb{R}$ converge. To do this we would first need to show that these sets of equivalence classes of Cauchy sequences (real numbers) are linearly ordered.

Unfortunately showing the linear ordering, while not especially hard, is time consuming. So we will again invoke the prerogatives of the teacher and brush all of the difficulties aside with the assertion that it is straightforward to show that the real numbers as we have constructed them in this section are linearly ordered and are complete. If you would like to see this construction in full rigor we recommend the book, *The Number System* by H. A. Thurston [16].

Comment. Thurston first builds $\mathbb{R}$ as we've indicated in this section. Then as a final remark he shows that the real numbers must be exactly the infinite decimals we saw in the previous section.

12.2.3 Dedekind Cuts

An advantage of building the reals via Cauchy sequences in the previous section is that once we've identified equivalent sequences with real numbers it is *very* clear how addition and multiplication should be defined.

On the other hand, before we can even start to understand that construction, we need a fairly strong sense of what it means for a sequence to converge and enough experience with sequences to be comfortable with the notion of a Cauchy sequence. Thus a good deal of high level mathematics must be mastered before we can even begin.

The method of "Dedekind cuts" first developed by Richard Dedekind (though he just called them "cuts") in his 1872 book, *Continuity and the Irrational Numbers* shares the advantage of the Cauchy sequence method in that, once the candidates for the real numbers have been identified, it is very clear how addition and multiplication should be defined. It is also straightforward to show that most of the field axioms are satisfied.

Comment. "Clear" does not mean "easy to do" as we will see.

In addition, Dedekind's method also has the advantage that very little mathematical knowledge is required to get started. This is intentional. In the preface to the first edition of his book, Dedekind states:

> This memoir can be understood by anyone possessing what is usually called common sense; no technical philosophic, or mathematical, knowledge is in the least degree required. (quoted in [5])

While he may have overstated his case a bit, it is clear that his intention was to argue from very simple first principles just as Euclid did.

His starting point was the observation we made in Chapter 2: The rational number line is full of holes. More precisely we can "cut" the rational line in two distinct ways:

1. We can pick a rational number, r. This choice divides all other rational numbers into two classes: Those greater than r and those less than r.

2. We can pick one of the holes in the rational number line. In this case *all* of the rationals fall into two classes: Those greater than the hole and those less.

But to speak of rational numbers as less than or greater than something that is *not* there is utter nonsense. We'll need a better (that is, a rigorous) definition.

As before we will develop an overall sense of this construction rather than a fully detailed presentation, as the latter would be far too long to include.

Our presentation will closely follow that of Edmund Landau's in his classic 1951 text *Foundations of Analysis* [7]. We do this so that if you choose to pursue this construction in more detail you will be able to follow Landau's presentation more easily.

Definition 12.2.12 ***Dedekind Cut***

A set of *positive* rational numbers is called a *cut* if

Comment. Take special notice that we are not using the negative rational numbers or zero to build our cuts. The reason for this will become clear shortly.

Property I It contains a positive rational number but does not contain all positive rational numbers.

Property II	Every positive rational number in the set is less than every positive rational number not in the set.
Property III	There is no element of the set which is greater than every other element of the set.

◊

Given their intended audiences, Dedekind and Landau shied away from using too much notation. However, we will include the following for those who are more comfortable with the symbolism as it may help provide more perspective. Specifically the properties defining a Dedekind cut α can be written as follows.

Property I	$\alpha \neq \emptyset$ and $\mathbb{Q}^+ - \alpha \neq \emptyset$.
Property II	If $x \in \alpha$ and $y \in \mathbb{Q}^+ - \alpha$, then $x < y$. (Alternatively, if $x \in \alpha$ and $y < x$, then $y \in \alpha$.)
Property III	If $x \in \alpha$, then $\exists\ z \in \alpha$ such that $x < z$.

Properties I-III really say that Dedekind cuts are bounded open intervals of rational numbers starting at 0. For example, $(0, 3) \cap \mathbb{Q}^+$ is a Dedekind cut (which will eventually be the real number 3). Likewise, $\{x|x^2 < 2\} \cap \mathbb{Q}^+$ is a Dedekind cut (which will eventually be the real number $\sqrt{2}$). Notice that care must be taken not to actually refer to irrational numbers in the properties as the purpose is to construct them from rational numbers, but it might help to ground you to anticipate what will happen.

Take particular notice of the following three facts:

1. Very little mathematical knowledge is required to understand this definition. We need to know what a set is, we need to know what a rational number is, and we need to know that given two positive rational numbers either they are equal or one is greater.
2. The language Landau uses is *very* precise. This is necessary in order to avoid such nonsense as trying to compare something with nothing like we did a couple of paragraphs up.
3. We are only using the *positive* rational numbers for our construction. The reason for this will become clear shortly. As a practical matter for now, this means that the cuts we have just defined will (eventually) correspond to the *positive* real numbers.

Definition 12.2.13 Let α and β be cuts. Then we say that α is less than β, and write

$$\alpha < \beta$$

if there is a rational number in β which is not in α. ◊

Note that, in light of what we said prior to Definition 12.2.13 (which is taken directly from Landau), we notice the following.

Theorem 12.2.14 *Let α and β be cuts. Then $\alpha < \beta$ if and only if $\alpha \subset \beta$.*

Problem 12.2.15 Prove Theorem 12.2.14 and use this to conclude that if α and β are cuts then exactly one of the following is true:

1. $\alpha = \beta$.
2. $\alpha < \beta$.
3. $\beta < \alpha$.

□

We will need first to define addition and multiplication for our cuts and eventually these will need to be extended to $\mathbb{R}$ (once the non-positive reals have also been constructed). It will be necessary to show that the extended definitions satisfy the field axioms. As you can see there is a lot to do.

As we did with Cauchy sequences and with infinite decimals, we will stop well short of the full construction. If you are interested in exploring the details of Dedekind's construction, Landau's book [7] is very thorough and was written with the explicit intention that it would be accessible to students. In his "Preface for the Teacher" he says

> I hope that I have written this book, after a preparation stretching over decades, in such a way that a normal student can read it in two days.

This may be stretching things. Give yourself at least a week and make sure you have nothing else to do that week.

Addition and multiplication are defined in the obvious way.

Definition 12.2.16 ***Addition on cuts***

Let α and β be cuts. We will denote the set $\{x + y | x \in \alpha, y \in \beta\}$ by $\alpha + \beta$. $\diamondsuit$

Definition 12.2.17 ***Multiplication on cuts***

Let α and β be cuts. We will denote the set $\{xy | x \in \alpha, y \in \beta\}$ by $\alpha\beta$ or $\alpha \cdot \beta$. $\diamondsuit$

If we are to have a hope that these objects will serve as our real numbers we must have closure with respect to addition and multiplication. We will show closure with respect to addition.

Theorem 12.2.18 ***Closure with Respect to Addition***

If α and β are cuts then $\alpha + \beta$ is a cut.

Proof. We need to show that the set $\alpha + \beta$ satisfies all three of the properties of a cut.

Property I Let x be *any* rational number in α and let x_1 be a rational number not in α. Then by Property II $x < x_1$.
Let y be *any* rational number in β and let y_1 be a rational number not in β. Then by Property II $y < y_1$.
Thus since $x + y$ represents a generic element of $\alpha + \beta$ and $x + y < x_1 + y_1$, it follows that $x_1 + y_1 \notin \alpha + \beta$.

Property II We will show that the contrapositive of ***Property II*** is true: If $x \in \alpha + \beta$ and $y < x$ then $y \in \alpha + \beta$.
First, let $x \in \alpha + \beta$. Then there are $x_\alpha \in \alpha$ and $x_\beta \in \beta$ such that $y < x = x_\alpha + x_\beta$. Therefore $\frac{y}{x_\alpha + x_\beta} < 1$, so that

$$x_\alpha \left(\frac{y}{x_\alpha + x_\beta} \right) < x_\alpha$$

and

$$x_\beta \left(\frac{y}{x_\alpha + x_\beta} \right) < x_\beta.$$

Therefore $x_\alpha \left(\frac{y}{x_\alpha + x_\beta} \right) \in \alpha$ and $x_\beta \left(\frac{y}{x_\alpha + x_\beta} \right) \in \beta$. Therefore

$$y = x_\alpha \left(\frac{y}{x_\alpha + x_\beta} \right) + x_\beta \left(\frac{y}{x_\alpha + x_\beta} \right) \in \alpha + \beta.$$

Property III Let $z \in \alpha + \beta$. We need to find $w > z$, $w \in \alpha + \beta$. Observe that for some $x \in \alpha$ and $y \in \beta$

$$z = x + y.$$

Since α is a cut, there is a rational number $x_1 \in \alpha$ such that $x_1 > x$. Take $w = x_1 + y \in \alpha + \beta$. Then

$$w = x_1 + y > x + y = z.$$

■

Problem 12.2.19 Show that if α and β are cuts then $\alpha \cdot \beta$ is also a cut. □

At this point we have built our cuts and we have defined addition and multiplication for cuts. However, as observed earlier the cuts we have will (very soon) correspond only to the positive real numbers. This may appear to be a problem but it really isn't because the non-positive real numbers can be defined in terms of the positives, that is, in terms of our cuts. We quote from Landau [7]:

> These cuts will henceforth be called the "positive numbers;" . . .
>
> We create a new number 0 (to be read "zero"), distinct from the positive numbers.
>
> We also create numbers which are distinct from the positive numbers as well as distinct from zero, and which we will call negative numbers, in such a way that to each ξ (I.e. to each positive number) we assign a negative number denoted by $-\xi$ ($-$ to be read "minus"). In this, $-\xi$ and $-\nu$ will be considered as the same number (as equal) if and only if ξ and ν are the same number.
>
> The totality consisting of all positive numbers, of 0, and of all negative numbers, will be called the real numbers.

Of course it is not nearly enough to simply postulate the existence of the non-negative real numbers.

All we have so far is a set of objects we're calling the real numbers. For some of them (the positive reals -- that is, the cuts) we have defined addition and multiplication. These definitions will eventually turn out to correspond to the addition and multiplication we are familiar with.

However we do not have either operation for our entire set of proposed real numbers. Before we do this we need first to define the *absolute value* of a real number. This is a concept you are very familiar with and you have probably seen the following definition: Let $\alpha \in \mathbb{R}$. Then

$$|\alpha| = \begin{cases} \alpha & \text{if } \alpha \geq 0, \\ -\alpha & \text{if } \alpha < 0. \end{cases}$$

Unfortunately we cannot use this definition because we do not yet have a linear ordering on $\mathbb{R}$ so the statement $\alpha \geq 0$ is meaningless. Indeed, it will be our definition of absolute value that orders the real numbers. We must be careful.

Notice that by definition a negative real number is denoted with the dash ('-') in front. That is χ is positive while $-\chi$ is negative. Thus if A is *any* real number then one of the following is true:

1. $A = \chi$ for some $\chi \in \mathbb{R}$ (A is positive)
2. $A = -\chi$ for some $\chi \in \mathbb{R}$ (A is negative)
3. $A = 0$.

We define absolute value as follows:

Definition 12.2.20 Let $A \in \mathbb{R}$ as above. Then

$$|A| = \begin{cases} \chi & \text{if } A = \chi \\ 0 & \text{if } A = 0 \\ \chi & \text{if } A = -\chi. \end{cases}$$

◊

With this definition in place it is possible to show that $\mathbb{R}$ is linearly ordered. We will not do this explicitly. Instead we will simply assume that the symbols "<" ">," and "=" have been defined and have all of the properties we have learned to expect from them.

We now extend our definitions of addition and multiplication from the positive real numbers (cuts) to all of them. Curiously, multiplication is the simpler of the two.

Definition 12.2.21 ***Multiplication***
Let $\alpha, \beta \in \mathbb{R}$. Then

$$\alpha \cdot \beta = \begin{cases} -|\alpha|\,|\beta| & \text{if } \alpha > 0, \beta < 0 \text{ or } \alpha < 0, \beta > 0, \\ |\alpha|\,|\beta| & \text{if } \alpha < 0, \beta < 0, \\ 0 & \text{if } \alpha = 0 \text{ or } \beta = 0. \end{cases}$$

◊

Notice that the case where α and β are both positive was already handled by Definition 12.2.17 because in that case they are both cuts.

Next we define addition.

Definition 12.2.22 ***Addition***

Let $\alpha, \beta \in \mathbb{R}$. Then

$$\alpha + \beta = \begin{cases} -(|\alpha| + |\beta|) & \text{if } \alpha < 0, \beta < 0 \\ |\alpha| - |\beta| & \text{if } \alpha > 0, \beta < 0, |\alpha| > |\beta| \\ 0 & \text{if } \alpha > 0, \beta < 0, |\alpha| = |\beta| \\ -(|\alpha| - |\beta|) & \text{if } \alpha > 0, \beta < 0, |\alpha| < |\beta| \\ \beta + \alpha & \text{if } \alpha < 0, \beta > 0 \\ \beta & \text{if } \alpha = 0 \\ \alpha & \text{if } \beta = 0 \end{cases}.$$

◊

Comment. Notice also that the fifth case refers to the addition as defined in the second case.

But wait! In the second and fourth cases of our definition we've actually defined addition in terms of subtraction. But we haven't defined subtraction yet! Oops!

This is handled with the definition below, but it illuminates very clearly the care that must be taken in these constructions. The real numbers are *so* familiar to us that it is extraordinarily easy to make unjustified assumptions.

Since the subtractions in the second and fourth cases above are done with positive numbers we only need to give meaning to the subtraction of cuts.

Definition 12.2.23 If α, β and δ are cuts then the expression

$$\alpha - \beta = \delta$$

is defined to mean

$$\alpha = \delta + \beta.$$

◊

Of course, there is the detail of showing that there is such a cut δ. (We warned you of the tediousness of all this.) Landau goes through the details of showing that such a cut exists. We will present an alternative by defining the cut $\alpha - \beta$ directly (assuming $\beta < \alpha$). To motivate this definition, consider something we are familiar with: $3 - 2 = 1$. In terms of cuts, we want to say that the open interval from 0 to 3 "minus" the open interval from 0 to 2 should give us the open interval from 0 to 1. Taking elements from $(0, 3)$ and subtracting elements from $(0, 2)$ won't do it as we would have differences such as $2.9 - .9 = 2$ which is not in the cut $(0, 1)$. A moment's thought tells us that what we need to do is take all the elements from $(0, 3)$ and subtract all the elements from $(2, \infty)$, restricting ourselves only to those which are positive rational numbers. This prompts the following definition.

Definition 12.2.24 Let α and β be cuts with $\beta < \alpha$. Define $\alpha - \beta$ as follows:

$$\alpha - \beta = \{x - y | x \in \alpha \text{ and } y \notin \beta\} \cap \mathbb{Q}^+.$$

◊

To show that, in fact, $\beta + (\alpha - \beta) = \alpha$, the following technical lemma will be helpful.

Lemma 12.2.25 *Let β be a cut, y and z be positive rational numbers not in β with $y < z$, and let $\varepsilon > 0$ be any rational number. Then there exist positive rational numbers r and s with $r \in \beta$, and $s \notin \beta$, such that $s < z$, and $s - r < \varepsilon$.*

Problem 12.2.26 Prove Lemma 12.2.25.

Hint.

Since β is a cut there exists $r_1 \in \beta$. Let $s_1 = y \notin \beta$. We know that $r_1 < s_1 < z$. Consider the midpoint $\frac{s_1+r_1}{2}$. If this is in β then relabel it as r_2 and relabel s_1 as s_2. If it is not in β then relabel it as s_2 and relabel r_1 as r_2, etc. □

Problem 12.2.27 Let α and β be cuts with $\beta < \alpha$. Prove that $\beta + (\alpha - \beta) = \alpha$.

Hint.

It is pretty straightforward to show that $\beta + (\alpha - \beta) \subseteq \alpha$. To show that $\alpha \subseteq \beta + (\alpha - \beta)$, we let $x \in \alpha$. Since $\beta < \alpha$, we have $y \in \alpha$ with $y \notin \beta$. We can assume without loss of generality that $x < y$. (Why?) Choose $z \in \alpha$ with $y < z$. By the Lemma 12.2.25, there exists positive rational numbers r and s with $r \in \beta$, $s \in \beta$, $s < z$, and $s - r < z - x$. Show that $x < r + (z - s)$. □

We will end by saying that no matter how you construct the real number system, there is really only one. More precisely we have the following theorem which we state without proof.

Comment. In fact, *not* proving this result seems to be standard in real analysis references. Most often it is simply stated as we do here.

Theorem 12.2.28 *Any complete, linearly ordered field is isomorphic to* $\mathbb{R}$.

Two linearly ordered number fields are said to be isomorphic if there is a one-to-one, onto mapping between them (such a mapping is called a bijection) which preserves addition, multiplication, and order. More precisely, if $\mathcal{F}_1$ and $\mathcal{F}_2$ are both linearly ordered fields, $x, y \in \mathcal{F}_1$ and $\phi : \mathcal{F}_1 \to \mathcal{F}_2$ is the mapping then

1. $\phi(x + y) = \phi(x) + \phi(y)$
2. $\phi(x \cdot y) = \phi(x) \cdot \phi(y)$
3. $x < y \Rightarrow \phi(x) < \phi(y)$.

Remember that we warned you that these constructions were fraught with technical details that are not necessarily illuminating. Nonetheless, at this point, you have everything you need to show that the set of all real numbers as defined above is linearly ordered and satisfies the Least Upper Bound property.

But we will stop here in order, to paraphrase Descartes, to leave for you the joy of further discovery.

Bibliography

[1] Robert E. Bradley and C. Edward Sandifer. *Cauchy's Cours d'analyse: An Annotated Translation.* Sources and Studies in the History of Mathematics and Physical Sciences. Springer, 2009.

[2] William Dunham. *Journey Through Genius.* Penguin Books, 1990.

[3] John Franks. Cantor's other proofs that $\mathbb{R}$ is uncountable. *Mathematics Magazine*, 83(4):283--289, October 2010.

[4] Judith Grabiner. *The Origins of Cauchy's Rigorous Calculus.* MIT Press, Cambridge MA, 1981.

[5] Stephen Hawking, editor. *God Created the Integers: The Mathematical Breakthroughs that Changed History.* Running Press, Philadelphia, London, 2005.

[6] H. Jahnke, editor. *A History of Analysis.* AMS Publications, Providence RI, 2003.

[7] Edmund Landau. *Foundations of Analysis.* Chelsea Publishing Company, New York, NY, 1966. Translated by F. Steinhardt, Columbia University.

[8] Thomas Levenson. *Newton and the Counterfeiter.* Houghton Mifflin Harcourt, 2009.

[9] Reviel Netz and William Noel. *The Archimedes Codex.* Da Capo Press, 2007.

[10] Isaac Newton. *Sir Isaac Newton's Two Treatises of the Quadrature of Curves and Analysis by Equation of an Infinite Number of Terms, Explained.* Society for the Encouragement of Learning, 1745. Translated from Latin by John Stewart, A. M. Professor of Mathematicks in the Marishal College and University of Aberdeen.

[11] J. J. O'Connor and E. F. Robertson. *The Brachistochrone Problem. http://www-gap.dcs.st-and.ac.uk", MacTutor History of Mathematics archive*

[12] Abraham Robinson. *Non-standard analysis.* North-Holland Pub. Co., 1974.

[13] Lucio Russo, translated by Silvio Levy. *The Forgotten Revolution: How Science Was Born in 300 BC and Why It Had to Be Reborn.* Springer, 1996.

[14] C. Edward Sandifer. *The Early Mathematics of Leonard Euler.* Spectrum, 2007. ISBN 10: 0883855593, ISBN 13: 978-0883855591.

[15] Dirk Struik, editor. *Source Book in Mathematics, 1200-1800.* Harvard University Press, Cambridge, MA, 1969.

[16] H. A. Thurston. *The Number System.* Blackie and Son Limited, London, Glassgow, 1956.

Index

Made in the USA
Monee, IL
02 November 2026

67384610R00103